DNA Beyond Genes

Vadim V. Demidov

DNA Beyond Genes

From Data Storage and Computing
to Nanobots, Nanomedicine,
and Nanoelectronics

 Springer

Vadim V. Demidov
Boston, MA, USA

ISBN 978-3-030-36436-6   ISBN 978-3-030-36434-2   (eBook)
https://doi.org/10.1007/978-3-030-36434-2

On the cover: Sketch of DNA-assisted fabrication of future nanometer-scale electronic circuits based on a self-assembled DNA network that serves, in turn, as a template for its subsequent functionalization with electronically operational elements (see Chapter 5.2 for details).

This Springer imprint is published by the registered company Springer Nature Switzerland AG
The registered company address is: Gewerbestrasse 11, 6330 Cham, Switzerland

To the love and passion for DNA, and to those readers who would be inspired by this amazing molecule of life, many new uses of which in a material world have yet to be discovered

Preface

In 1979, Francis Crick, prominent British scientist and co-discoverer of the DNA double-helical structure, wrote [1]: "Gradually DNA became better known. Paul Doty [1] told me that shortly after lapel buttons came in he was in New York and to his astonishment saw one with 'DNA' written on it. Thinking it must refer to something else, he asked the vendor what it meant. '*Get with it, bud,*' the man replied in a strong New York accent, '*dat's the gene.*'" Indeed, nowadays almost everybody knows what our chromosomes are made of: a few years ago, 93% of people correctly said during public survey that gene is a piece of DNA [2]. And we all are well informed that the DNA's genetic nature brought us amazing things, such as recombinant proteins, DNA vaccines, gene therapy, and animal cloning.

But beyond the role as a genetic information carrier, DNA has recently attracted a great deal of attention from physicists, chemists, and materials scientists for its totally unique physicochemical properties and abilities as a material molecule that may bring us new smart and robust tools and technologies. However, such a momentous side of DNA is less known to general audience, which strongly motivated me to write this book for a wide readership. In it, I will open to the interested readers a fascinating whole new world of "uncommon" DNA—the material world, which embraces the emerging fields of DNA electronics, structural DNA nanotechnology, DNA computing and DNA data storage, and the rising realm of DNA machines[2].

These exciting developments promise to further revolutionize our life. By presenting them to the readers, I hope that my writing would be interesting to those, who would like to extend their educational background (frankly, I greatly enriched myself with a new knowledge while working with the materials for this book, even despite my many years of involvement into DNA biotech research). I also hope to inspire the young generation of future scientists and technologists for new breakthroughs in the DNA material science and technologies.

The fields of employing DNA in a material world are now burgeoning, so the reader of this book will discover that there is no lack of thrilling surprises, turns, and

---

[1] Paul Doty was Mallinckrodt Professor of Biochemistry at Harvard University, specializing in the characterization of biopolymers such as DNA, proteins, and collagen by optical methods.

[2] Though several specialized books have appeared in recent years dealing with different topics related to the contents of this book, to my knowledge this is the first book portraying many fields of DNA in a material world to a wide readership altogether in a single volume.

twists in the continuing development of "uncommon" DNA. I am certain that further promising developments in the area of novel DNA-based materials will be surfaced soon, which are not covered by this book. Yet, my intention was not to write a comprehensive and thoroughly updated review of the field, but to present my readers with the wealth of ideas floating out there. I decided this in the interests of the concept of my book by realizing the vast and rapidly extending space of current studies on material properties of DNA, as well as by keeping in mind the wise words once said by Kozma Prutkov [3]: "Nobody can embrace the immense!"[3,4]

Boston, MA, USA   Vadim V. Demidov
August 2019

References

1. Crick F (1979) How to live with a golden helix. The Sciences 19:6–9
2. Haga SB et al (2013) Public knowledge of and attitudes toward genetics and genetic testing. Genet Test Mol Biomarkers 17:327–335
3. Prutkov K (1854) The fruits of reflection: thoughts and aphorisms [translated by George Gibain. In: Portable nineteenth century Russian reader (1993). Penguin Putnam Trade, New York, NY]
4. Yashon R, Cummings MR (2019) DNA forensics. Momentum Press
5. The New York Times Editorial Staff (ed) (2018) DNA testing: genealogy and forensics (looking forward). New York Times Educational Publishing, New York
6. Martin-Fernandez B et al (2017) Electrochemical genosensors in food safety assessment. Crit Rev Food Sci Nutr 57:2758–2774
7. Datta M et al (2017) Gene specific DNA sensors for diagnosis of pathogenic infections. Indian J Microbiol 57:139–147
8. Kokkokoglu F, Senel M (2013) DNA biosensors: detection of breast cancer from genetic mutations. LAP Lambert Academic Publishing
9. Ozsoz M (ed) (2012) Electrochemical DNA biosensors. Jenny Stanford Publishing

---

[3] Fictional author Kozma Prutkov is a collective pen name of the four distinguished Russian writers and poets of nineteenth century, Aleksey Tolstoy and three Zhemchuzhnikov brothers. Under this pseudonym, they published numerous immensely popular aphorisms essentially about everything, some of which serve as epigraphs to certain sections of this book.

[4] I decided to not cover in this book the DNA diagnostics/forensics and DNA-based sensors since, though these topics are somehow related to DNA uses in a material world, they both rely on various approaches for detection of specific parts of genomic DNA. Anyway, several nice review articles and books on DNA testing and DNA biosensors have recently been published to fill this gap for the interested reader [4–9].

Keeping up with the directions and applications of DNA is a never-ending job.
I. Edward Alcamo, DNA Technology: The Awesome Skill (*2000*)

The exploitation of DNA for material purposes presents a new chapter in the history of the molecule.
Nadrian C. Seeman, DNA in a Material World (*2003*)

Advance Praise

Most students of DNA, and lay readers as well, are interested in the absolutely essential role it plays in biology. However, the properties which make DNA the carrier of genetic information also make it an extraordinary material that can be used as the backbone for a wide variety of nanoengineering applications—these range from information storage and computation to molecular machines and devices to artfully designed logos and symbols. The perfect self-recognition of DNA sequences makes it an ideal building block to synthesize more and more elaborate constructions, and imaginative scientists have probably only just scratched the surface of what can eventually be created. Here for the first time in this wonderful book Vadim Demidov explores the full range of the non-biological applications of DNA.

Charles R. Cantor
Biomedical Engineering, Boston University, Boston, MA, USA
USA National Academy of Sciences, Washington, DC, USA

Great update on current research involving "uncommon" DNA, fun and inspiring to read it!

Anastasia Yaroslavsky
Biomedical Engineering, Boston University, Boston, MA, USA

While I tend to think of DNA only in terms of genetic material, this book clearly presents prospective multiple uses of DNA related to nanoelectronics, data storage, smart materials, and molecular machines first time collected into a single volume. It is a great resource to quickly and easily expand my understanding of DNA as a multipurpose matter, and I would expect the book to do the same for other readers.

Kennyn Statler, Ph.D.
Management Consulting, ipCapital Group, Williston, VT, USA

Fascinating and insightful overview of creative DNA applications beyond genome and related uses, some of which may seem straight out of science fiction pages. I particularly enjoyed the sections on DNA cryptography and nanobots. For such a complicated subject, the book is surprisingly jargon-free and will appeal to a broad audience of experts and novices, including even interested laypersons.

Saheli Sarkar
Department of Pharmaceutical Sciences, Northeastern
University, Boston, MA, USA

The author did a good job of introducing DNA and related fields to the reader, who may be unfamiliar with the internal workings of the subject. Once introduced he creates a story, allowing the reader to follow along effortlessly. The referencing at the end of chapters is a useful tactic for eager readers to continue reading more about a given topic. A major strength of this book is the ideas stimulated in the reader. This book is not trying to explain the precise mechanisms of action, but it is simply trying to enlighten the reader about the cutting-edge uses of DNA being applied today. Outside-the-box thinking is the cornerstone to scientific break-throughs, and this book allows anyone who reads it to glimpse at some of the potentials of DNA and what is yet to come, such as using DNA nanobots to sort items on a microscopic level. It is common that science fiction becomes reality due to scientific progress, and this book nicely illustrates a few examples.

Shaun Filliaux, Ph.D.
University of Nebraska Medical Center, Omaha, NE, USA

I believe that this book will not only inspire young scientists out there to come up with more breakthrough inventions employing "uncommon" DNA, but it will also engage and inform the curious common reader like me about the many uses of DNA related to everyday areas such as drug delivery, identity tags, electronics, and computing (and even for miniscule painting and sculpting!). This book is nicely illustrated, which makes it very readable, and I recommend it to anyone who wants to learn more about gene molecules in a material world.

Elif Burduroglu
Bostonian artist

Abbreviations

1D	One-dimensional
2D	Two-dimensional
3D	Three-dimensional
A	Adenine
ASU	Arizona State University
C	Cytosine
Caltech	California Institute of Technology
CD	Compact disc
CNT	Carbon nanotube
DMF	Digital microfluidic
DNA	Deoxyribonucleic acid
DNAzyme	DNA enzyme
dsDNA	Double-stranded DNA
DVD	Digital versatile disc
EPR	Enhanced permeability and retention
FRET	Fluorescence resonance energy transfer
G	Guanine
GDa	Gigadalton
iDNA	Informational DNA
kb	Kilobase or kilobase pairs
MDa	Megadalton
mRNA	Messenger RNA
NIH	National Institutes of Health
NYU	New York University
PCR	Polymerase chain reaction
PNA	Peptide nucleic acid
PPV	Polyphenylene vinylene
RCA	Rolling-circle amplification
RNA	Ribonucleic acid
siRNA	Small interfering RNA
ssDNA	Single-stranded DNA
SST	Single-stranded tile
T	Thymine

Acknowledgements

I would like to thank all researchers who kindly provided me with figures and micrographs from their exciting studies. I am also grateful to the team at Springer Nature for bringing this book to life. Especially, I wish to express sincere gratitude to my editor Merry Stuber for her professional advices and assistance in polishing this manuscript, and to Deepak Ravi and Kala Palanisamy who oversaw the production of this book. Besides, I gratefully acknowledge the help of anonymous reviewers of my draft manuscript for useful comments and suggestions.

Contents

About the Author

Vadim V. Demidov, PhD is a Senior Analyst in the Biotechnology and Pharmaceuticals Group at Global Prior Art, Inc. (Boston, USA), an intellectual property research and analysis firm.

He received MS degree in Physical/Chemical Engineering from Moscow Institute of Physics and Technology (MIPT, a leading Russian technical university, *aka* "the Russian MIT") and PhD degree in Biophysics from the Institute of Molecular Genetics of the Russian Academy of Sciences (IMGRAS) and MIPT. Before joining the Global Prior Art company in 2008, he has worked for almost 30 years in academia and industry worldwide, serving lately as a research professor and senior scientist at prestigious institutions, such as Moscow Institute of Biotechnology and Institute of Molecular Genetics (Russia), Copenhagen University (Denmark), George Mason University and Boston University (USA).

Vadim Demidov is well known in molecular biology and biotechnology fields for his innovative studies related to peptide nucleic acid (PNA). He has pioneered the use of bis-PNA openers for formation of the so-called PD-loops and other related DNA nanostructures, and employed pseudocomplementary PNA benders for assembly of DNA minicircles.

During his research career, Demidov has published over 50 peer-reviewed research papers. In addition, he holds several US and international patents on nucleic acids biotechnology and environmental monitoring. He has presented at many international conferences, wrote several review articles and book chapters, and published two books—*DNA Amplification: Current Technologies and Applications* (Horizon Bioscience, 2004) and *Rolling Circle Amplification (RCA): Toward New Clinical Diagnostics and Therapeutics* (Springer, 2016).

Besides the scientific research, Demidov is an active freelance writer; he wrote numerous news, commentary, and feature articles for *Modern Drug Discovery* (American Chemical Society), *Drug Discovery Today* (Elsevier), *Trends in Biotechnology* (Elsevier), *Drug Discovery & Development* (Reed-Elsevier), and *Chemistry & Industry* (Society of Chemical Industry, UK). He also served as the Editorial Board member for *Trends in Biotechnology*, *Expert Review of Molecular Diagnostics*, *Expert Opinion on Medical Diagnostics*, *Current Medicinal Chemistry,* and *Open Medicinal Chemistry*.

Dr. Demidov was awarded Silver Medal from All-Union National Exhibition of Economic Achievements (Moscow, USSR, 1988) and Medal of Honor from International Biographical Centre (Cambridge, UK, 2007). He is listed in several international biographical directories, including *Who's Who in Science and Engineering*, *Who's Who in Medicine and Healthcare*, *Who's Who in America*, *Who's Who in the World,* and *Dictionary of International Biography*.

Away from the office, Vadim Demidov likes biking Boston, ocean swimming, ice skating, and traveling the world with his daughter Julia (the photo above was taken by her in 2017 in Swiss Alps near Lucerne). He is also a devoted pet lover, and shares with his wife Inna a home with two shelter rescue chihuahuas, Peanut and Lola, numerous birds, and other small animals.

Introduction: DNA Basics—A Primer on DNA

1

All rising to great places is by a winding stair.
Francis Bacon *(English philosopher),* Of Great Place; Essays, or Counsels Civil and Moral *(1625)*

The DNA model of Watson and Crick looks like a diamond as big as the Ritz.
Maxim D. Frank-Kamenetskii *(Russian-American biophysical theoretician),* Unraveling DNA: The Most Important Molecule of Life *(1997)*

To better understand the contents of this book, the lay reader needs to know some basic facts about DNA, and the aim of the following brief introduction is to provide such a necessary background. The reader may also bump in the next chapters into some unusual terms so the Glossary at the end of this book should help him/her to understand the meaning of such words. And those readers who are experienced enough in this subject could skip this chapter.

DNA, or *d*eoxyribo*n*ucleic *a*cid, is a natural polymeric molecule, the exceptional structure of which encodes the genetic instructions directing the development and functioning of every single live cells and more complex organisms living on our planet. As seen in Fig. 1.1, the 3D structure of DNA in its major natural form, B-DNA, looks like a right-handed spiral-shaped ladder or staircase constructed of the two helical side strands, called DNA backbones, and serves as a kind of rails, which are joined at regular intervals by horizontal flat steps, called base pairs.

The DNA backbones are made up of alternating phosphate and sugar residues joined by ester bonds, whereas the base pairs in a double-stranded DNA molecule are the couples of complementary nitrogenous bases (aka nucleobases), consisting of a purine residue in one strand linked by hydrogen bonds to a pyrimidine residue in the other (see Fig. 1.2). There are four nucleobases in DNA, two purines (adenine and guanine) and two pyrimidines (cytosine and thymine), which form two different

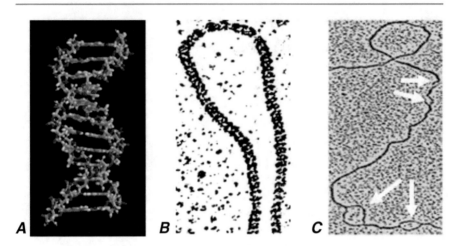

Fig. 1.1 DNA in pictures. (**a**) The 3D model of a DNA double helix in its major natural form called B-DNA. This model is built of small colored balls representing the hydrogen, oxygen, nitrogen, carbon, and phosphorus atoms of which DNA is made, and it shows the complementary pairs of nucleobases that lie horizontally between the two spiraling side strands made of a chain of sugar–phosphate residues that carry the base pairs. Note that the helix here is right handed, as it is advanced by turning clockwise. (**b**) The micrograph of a small section of the double-helical DNA taken with an electron microscope; the DNA spiral structure is well seen here. In obtaining this picture, a beam of electrons was used instead of photons (the particles of light) to visualize the molecule because electrons can provide with larger magnification and better resolution. (**c**) Electron micrograph of a larger section of the double-helical DNA, showing its threadlike filamentous form. This picture was taken at elevated temperatures, so that pointed by arrows loops correspond to the melted (denatured) DNA regions, where the two DNA strands are separated from each other by heat. (Courtesy of late Ross Inman, University of Wisconsin-Madison)

base pairs: cytosine (C) always pairs with guanine (G) to make GC base pair, and adenine (A) with thymine (T) to make AT base pair.

The two backbone strands of the DNA double helix are antiparallel to each other in a sense that some chemical bonds between the backbone units are asymmetric, therefore defining some direction so that these two strands are running in opposite directions, as it is shown by red arrows in Fig. 1.2 (see the entry "3'-end and 5'-end" in Glossary for more details). The four different A, C, G, and T nucleobases are attached to each sugar of the backbone, and namely their unique order along the backbone, called DNA sequence, encodes genetic information.

The rules of A↔T and G↔C complementarity allow to precisely replicate any particular DNA sequence encoded by either of DNA strands; these rules also direct a self-assembly of a pair of single complementary matching DNA strands into the double helix. This structure of DNA molecule, including specific pairing of nucleobases (called Watson-Crick base pairs) and general principle of DNA replication, was discovered by Watson and Crick almost 60 years ago [1]. Now we know that the DNA sequence is read in live cells by using the genetic code, which specifies the sequence of the amino acids within proteins. DNA transcribes genetic messages onto related nucleic acid, RNA (*ribo*nucleic *a*cid), in a process called transcription. Finally, the genetic code instructed by a particular DNA sequence and transcribed into an RNA molecule is transformed into a specific protein by a process called translation [2, 3].

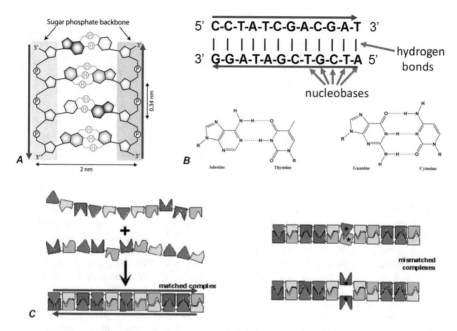

Fig. 1.2 DNA in schematics. (**a**) Diagram of DNA chemical structure. A double-helical DNA molecule is composed of two strands, each of them being a polymer of basic structural units called nucleotides. A nucleotide is composed of one of the four nucleobases (A, T, G, or C) linked to a pentose sugar residue (called deoxyribose), which in turn is linked to a phosphate residue. Red arrows here and in the two other schematics indicate the antiparallel polarity of DNA strands in a hydrogen-bonded molecular duplex. (**b**) Schematics of individual common AT and GC base pairs and those in a double-stranded DNA molecule. (**c**) Representation of short DNA single strands called oligonucleotides as elastic array of four primary elements, i.e., A, T, G, and C nucleobases, with T (blue element) matching only A (red element) and vice versa, and C (yellow element) matching only G (green element) and vice versa. The two fully complementary DNA single strands bind together in the annealing process forming strong and specific double-stranded complex, as it is shown on the left. When oligonucleotides are not fully complementary, as it is shown on the right, even single mismatches (the mismatched pairs are marked by asterisks) lead to formation of incorrect complexes, which are much less stable and do not form under appropriate conditions

The double-helical DNA can also adopt a number of alternative conformations [3, 4], two of which, called A-DNA and Z-DNA, are biologically active ones and play the roles in this book, too. Moreover, the double helix is not the only known form of DNA: uncommon, non-Watson-Crick base-pairing[1] allows the third DNA strands to wind around the regular DNA duplexes to form the triple-stranded helices called DNA triplexes [4, 5]. Besides, the non-Watson-Crick base-pairing facilitates formation of quadruple helical forms of G-rich single-stranded DNA called G-quadruplexes [6] and featuring the planar quartets of Hoogsteen-bonded guanines (see Fig. 1.3 for these alternative DNA helical structures).

---

[1] In addition to common Watson-Crick AT and GC base pairs, DNA nucleobases can form several unconventional pairs, including those with alternative hydrogen-bond pairing of A to T, and G to C (known as Hoogsteen base pairs), as well as AA, GG, and AG base pairs.

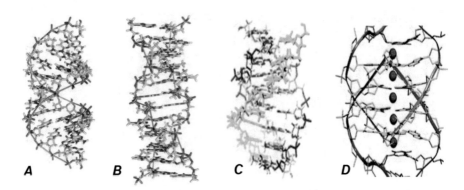

Fig. 1.3 Alternative DNA helical structures. (**a**) The 3D model of A-DNA—one of the other possible double-helical DNA structures—which is formed from B-DNA under dehydrating conditions when the concentration of water molecules surrounding DNA is rather low. A-DNA is a right-handed double helix, like the more common B-DNA, but it has a more compact helical structure whose base pairs are not perpendicular to the helix axis as in B-DNA. (**b**) The model of Z-DNA, which is formed by alternating sequences of purine-pyrimidine nucleotides in the presence of bivalent salts or at high concentration of monovalent salts. It is a left-handed double-helical DNA structure wherein the sugar–phosphate backbone has a zigzag pattern, unlike smooth winding of DNA backbone in right-handed A- and B-forms. (**c**) The model of a DNA triplex, which is formed when B-like DNA duplex of special sequence (i.e., homopurine-homopyrimidine stretches), binds a third DNA strand (shown in green) via additional hydrogen bonds between corresponding nucleobases, thus forming triplet base pairs. (**d**) The model of a DNA quadruplex, which is formed by G-rich DNA sequences when the four DNA strands adopt right-handed helical structures with hydrogen-bonded flat tetrads of G nucleobases, stabilized by metal ions located at the tetrad's center (shown as red balls)

Electron microscope makes it possible to look at the shape of individual DNA molecules. The spiral DNA structure is well seen in Fig. 1.1b. Figure 1.1c reveals the DNA as a long polymer, called a polynucleotide (i.e., a polymer of nucleotide units), which has a threadlike filamentous form, with internal loops being formed by melted (denatured) DNA regions, where the two DNA strands are separated from each other, being pulled apart like when you unzip a closed zipper. Importantly, short DNA duplexes behave like the stiff rods, since duplex DNA flexibility becomes noticeable only at lengths well over of 150 base pairs.

In contrast, single-stranded DNAs are much more flexible and this feature allows the DNA strands to form a variety of other non-B DNA conformations, like cruciform, multi-way junctions, and intramolecular triplexes and quadruplexes, which makes it possible to employ DNA as a clay to sculpt various nano-shapes and nano-constructs, and also to built nanobots and nanomachines. It is worth to mention here as well that due to phosphate residues, the surface of the DNA double helix is tightly covered with negative charges. This feature attracts metal cations to negatively charged polyanionic DNA, thus allowing the DNA filament to be easily "metalized."

As it will be shown in this book, despite the fact that the main role of DNA molecules in vivo is the long-term storage and replication of genetic information, the unique DNA molecular and supramolecular structures, their potential for self-assembly, and the DNA ability of conformational changes in response to external stimuli all these make DNA a smart material[2] and suggest some other important DNA uses beyond the gene concept and biologically related goals, which are presented in the next chapters covering the emerging fields of DNA in the world of materials from data storage and processing to nanobots, nanomedicine, and nanoelectronics. But as I already noted this in Preface, the DNA diagnostics/forensics applications and DNA-based sensors are deliberately left without consideration in this book: though these topics are somehow related to uses of DNA in a material world, they still rely on various approaches for detecting the DNA of living things. Therefore, I refer the interested reader to recently published review articles and books on DNA testing and DNA biosensors referenced in Preface.

For certain applications described in this book, DNA needs an assistance of so-called DNA-processing enzymes. These are highly specialized and accurate molecular instruments (or molecular machines) comprising specific proteins that perform a variety of basic functions with DNA filaments, including the precise copying of DNA strands (polymerases), cutting the DNA strands into smaller pieces (restriction enzymes), or otherwise joining the DNA pieces to get a bigger chunk of DNA (ligases).

I also would like to highlight here one more side of "uncommon" DNA which I have discovered while writing the editorial for *Trends in Biotechnology* journal on 50th anniversary of the double helix discovery [7]. To my surprise, I have found that in addition to science and technology DNA penetrates essentially all corners of our everyday life, and has become a cultural icon and even a commodity. The double helix is such a striking symbol that it can be seen everywhere—on commercial posters and postage stamps, on T-shirts and mugs, and even in artworks or architecture, and in the form of perfume bottles (see Fig. 1.4 and ref. [8]). There is no doubt for me that the twenty-first century will be a new era for DNA—the molecular quintessence of life—now in a material world.

---

[2] Smart materials are a category of multifunctional materials with physical or chemical properties that can be controllably altered in reaction to certain exterior factors, such as moisture, temperature, pH, electric or magnetic fields, light, or chemical compounds.

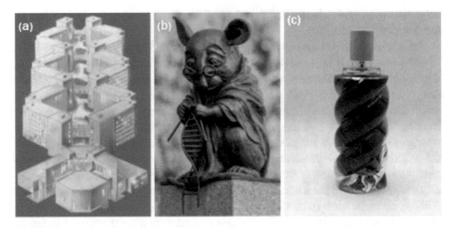

Fig. 1.4 Inspired by DNA. (**a**) The top view of a unique complex of buildings, the Moscow Institute of Bioorganic Chemistry, designed to resemble the DNA double helix. This institute is one of the world's leading centers in various areas of life sciences, including biotechnology. (**b**) The "*Monument to the laboratory mouse*" designed by artist Andrew Kharkevich and installed near the Institute of Cytology and Genetics in Novosibirsk, Russia, to commemorate the sacrifice of numerous mice in genetic research; the mouse is knitting a symbolic DNA double helix. (**c**) Spray bottle for perfume "DNA by Bijan," which was launched in 1993 by late Iranian designer of menswear and fragrances Bijan Pakzad

References

1. Watson JD, Crick FHC (1953) Molecular structure of nucleic acids: a structure for deoxyribose nucleic acid. Nature 171:737–738
2. Allison LA (2007) Fundamental molecular biology. Wiley-Blackwell, Hoboken, NJ
3. Sinden RR (1994) DNA structure and function. Academic, San Diego
4. Frank-Kamenetskii MD (1997) Unraveling DNA: the most important molecule of life. Addison-Wesley, New York
5. Soyfer VN, Potaman VN (1995) Triple-helical nucleic acids. Springer, New York
6. Neidle S, Balasubramanian S (2006) Quadruplex nucleic acids. RSC Publishing, Cambridge
7. Demidov VV (2003) Golden jubilee of the DNA double helix. Trends Biotechnol 21:139–140
8. Cellania M (2014) 8 DNA sculptures from around the world. http://mentalfloss.com/article/54579/8-dna-sculptures-around-world

Hiding and Storing Messages and Data in DNA

2

In the living cell, DNA functions as a principal informational molecule: it holds a linear array of heritable genetic information and it is the storehouse of this information. And while in nature DNA encodes proteins, the current ability to readily generate sufficiently long stretches of synthetic DNAs can be used for encrypting in DNA sequences some secret messages and for encoding and storing in DNA the large amounts of digital data, or even for tagging (or marking) and tracing various objects and materials with the coded DNA labels. To practically do so, we also need the ability to selectively amplify tiny amounts of DNA carrying an encoded or encrypted message or a label, and to read the DNA sequence in order to retrieve the DNA message (or DNA label). Fortunately, all this can now be done automatically by special machines.

Computer-controlled machine for DNA synthesis is called the DNA synthesizer (Fig. 2.1a). The process for DNA synthesis has been fully automated since the late 1970s and it is based on a sequential addition of the four DNA letters (known as nucleotides) to a growing DNA strand attached to a special bead. For the machine to know which nucleotide to add at each step, the operator enters the desired sequence in a computer. Currently, up to about 200-nucleotide-long synthetic DNAs can be obtained this way, which is quite enough to use them for encoding short messages and labels.

Up to millions of copies of a particular DNA sequence can be obtained by using another programmable machine designed in late 1980s which is called a thermocycler or PCR machine. This device amplifies a specific sequence of DNA via the polymerase chain reaction (PCR; see Glossary for PCR basics), and it has a thermal block with holes where tubes holding the amplification reaction mixtures can be inserted. A thermocycler then raises and lowers the temperature of the block in discrete, pre-programmed steps, and during each thermal cycle lasting just a few minutes the number of DNA molecules doubles. As a result, after 20 cycles, this would give approximately a millionfold amplification in about an hour or so. The original thermocycler was quite a bulky machine, but current thermocyclers are becoming smaller and smaller, and mini-PCR machine with the size of a smartphone has recently been designed (Fig. 2.1b).

© The Author(s), under exclusive licence to Springer Nature Switzerland AG 2020
V. V. Demidov, *DNA Beyond Genes*,
https://doi.org/10.1007/978-3-030-36434-2_2

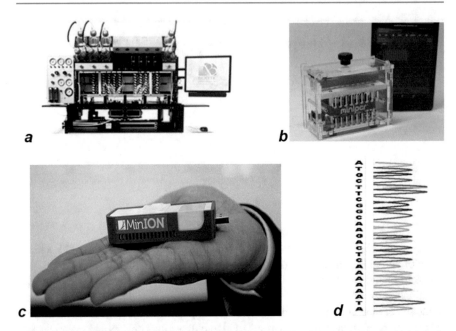

Fig. 2.1 Examples of machines used for DNA synthesis, amplification, and sequencing. (**a**) A typical view of an automated DNA synthesizer; red bottles at the top are for reagents and products; display of computer controlling all operations is seen on the right. (**b**) Miniature thermocycler recently developed by miniPCR, a Cambridge (MA)-based company and used for DNA amplification in the International Space Station. The smartphone seen behind this small device is shown as a scale of its largest dimension—around 5 in. (13 cm), and it is downloaded with apps to control the thermocycler and analyze amplification data. (**c**) Miniature DNA sequencer recently developed by Oxford Nanopore Technologies, a UK-based company, that measures a nucleotide-specific change in the current through a nanoscale hole (called nanopore) when a DNA strand passes through it. (**d**) An example of the results of automated DNA sequencing: each color on the so-called sequenogram corresponds to a specific nucleotide, which can be readily converted to a sequence read

The first automated machine for DNA sequencing, a DNA sequencer, was also designed in late 1980s. Since then, two more new generations of DNA sequencers, called as next generation and third generation, were built based on different principles of operation, making these machines more robust and more compact. Figure 2.1c shows the smallest DNA sequencer currently available, called MinION, which has the size smaller than a hand palm. Given a sample of DNA, a DNA sequencer is used to determine the order of the four DNA letters, A, C, G, and T, representing the four nucleotide bases of a DNA strand—adenine (A), cytosine (C), guanine (G), and thymine (T)—covalently linked to a phosphodiester backbone. This is then reported as a text string, called a sequence read or sequenogram (see Fig. 2.1d). The third-generation DNA sequencers read the nucleotide sequence of a single DNA molecule with high speed in real time, instantly displaying sequencing data on the laptop's screen as they are generated [1]. As such, these portable devices obtain a billion-nucleotide-long sequence in about 10–20 min [1, 2] so that the entire human genome, which is three billion nucleotides long, will be sequenced in an hour or less!

These three smart devices are all we need now to hide, to store, and to retrieve various information in DNA molecules.

2.1   DNA Cryptography

Since ancient times, cryptography is known as an ingenious art of hiding secret messages. The purpose of encryption is to transform important message into kind of "innocent" form for keeping it on the hush from others. Remarkably, an invisible smear of synthetic DNA can be used as a carrier and a keeper of highly sensitive and strictly confidential information.

To conceal such information, e.g., to encode a secret message in DNA, we need to establish an encryption code. And the easiest way it can be done is to use a simple substitution cipher, i.e., to encode usual text letters and other text symbols by triplets or quadruplets of the four DNA letters. For encoding of English-language texts, we need to encrypt all 26 letters of English alphabet and a space (to separate words), plus 10 common punctuation marks (period, comma, question mark, exclamation point, apostrophe, quotation mark, hyphen, colon, semicolon, and parentheses), as well as 10 digits: 0, 1, 2, 3, 4, 5, 6, 7, 8, and 9.

Altogether, we need to encrypt just 47 characters for encoding messages in English, whereas triplets of the 4 DNA letters sufficiently provide us with $4 \times 4 \times 4 = 64$ different 3-letter combinations (evidently, the doublet code yields only $4 \times 4 = 16$ different combinations of 4 letters, which is not sufficient for complete encoding of even plain texts). For basic Japanese writing systems, known as *hiragana* and *katakana*, which have 46 characters (or 71 with diacritical marks) and a dozen commonly used punctuation marks, the quadruplets of DNA letters, providing with $4 \times 4 \times 4 \times 4 = 256$ different four-letter combinations, can be used to code all 70+ Japanese alphanumeric characters.

An example of a triplet DNA code for encoding in English is shown in Table 2.1. By using this coding table, we can compose the DNA sequence encoding the hypothetical message from James Bond: "*No. 1 is ESB, 007*" (where ESB stands for *E*rnst *S*tavro *B*lofeld, a supervillain from the James Bond series of novels and films, who is the head of the global criminal organization SPECTRE). The corresponding 51-nucleotide-long DNA sequence is AGC ACC GGT GTA CCA GAC AAC ATT GTA GTT ATT AAA GAG GAC GGA TAT TGG.

To retrieve this message from the DNA sequence, we will need to amplify the DNA fragment comprising it (naturally, the DNA carrying a secret message should be in microscopic amounts to easily conceal it). For doing this, we need to attach to both sides of the message the sequences for amplification primers (see Fig. 2.2 and Glossary for primer term). Typical length of primers for PCR amplification is about 20 nucleotides so that the entire amplifiable DNA construct carrying a secret message will be no longer than 100 nucleotides, which could be readily produced by currently available DNA synthesizers.

Then, the intended recipient, who should be aware of both the "secret-message" DNA PCR primer sequences and the encryption code, could readily amplify the

Table 2.1 Representative DNA encryption code

A = AAT	B = AAA	C = TTT	D = CTT	E = GTT	F = GAA	G = ATG
H = TAA	I = AAC	J = AAG	K = ACG	L = AGG	M = ATC	N = AGC
O = ACC	P = CCC	Q = CAT	R = TCC	S = ATT	T = GCC	U = CAA
V = ACT	W = AGT	X = CAC	Y = ATA	Z = CTC	= GTA or GAC[a]	
1 = CCA	2 = CGC	3 = GGG	4 = TTG	5 = CCT	6 = CCG	
7 = TGG	8 = CGG	9 = AGA	0 = GGA or TAT[a]			
. = GGT	, = GAG	? = GTG	! = TCA	' = GCA	" = GCT	- = TCG
: = GAT	; = ACA	() = GCG				

[a]Since for encoding in English this triplet code is excessive (64 different triplets are available in total for encoding only 47 characters), some characters can be encoded by more than one triplet, therefore making code more complicated

"messenger" DNA, and after reading the DNA sequence decode the message. Importantly, amplification primers provide with the additional level of secrecy: it would be practically impossible to amplify (and therefore to read) the message without knowing the specific primer sequences since more than billions of billion combinations of two random primers should be tested to find the right one! And if necessary, a "secret message" DNA could even be more hidden in a very complex background of unrelated, e.g., human genomic DNA because of the power of PCR to selectively amplify just a few copies of specific DNA mixed with millionfold excess of physically similar DNA strands.

To test the practicality of this technique, its inventors, the researchers from Mount Sinai School of Medicine in New York, encoded, by using encryption code different but similar to that shown in Table 2.1, in a 69-nucleotide-long DNA sequence short text, containing probably one of the most famous WWII secrets: "*June 6 invasion: Normandy.*" Then, they diluted a few copies of this DNA "message" into a solution of short fragments of total human DNA and pipetted a drop of this solution over a period printed on a paper to mark its location. After excision of a small piece of paper with the printed punctuation mark, now carrying the secret message, and placing it in the microtube for PCR amplification and subsequent sequencing, the scientists were able to decode the message they encoded in DNA [3].

Fig. 2.2 Amplifiable DNA construct carrying a secret message. 5′ and 3′ denote the termini of a DNA strand according to common convention in nucleic acid chemistry that polynucleotide sequences are written in the 5′ → 3′ direction. The DNA polymerase will start replication at the 3′-end of the primer 2 after its binding to the corresponding complement sequence at the 3′ termini of the coded message. Then, primer 1 will be employed to generate a second copy, and so on

Moreover, they mailed to themselves the printed letter with the 100 copies of a DNA "message" attached to some known place in this letter. When this letter arrived, the scientists extracted the DNA, amplified the strand containing the

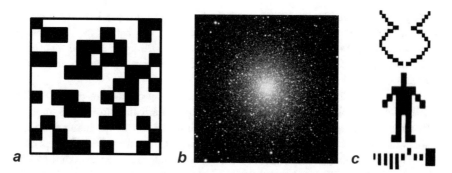

Fig. 2.3 Communicating with aliens. (**a**) Exemplary pictogram constructed from ϕX174 DNA by decoding its nucleotide sequence using various possible decryption algorithms (adapted from [4]). It looks like a 2D barcode commonly used now for marking small items, like printed labels and letters. (**b**) Telescopic view of M13, the globular star cluster in Hercules constellation, to which the Arecibo radio message was sent in 1974. (**c**) Parts of the encoded Arecibo message: schematics of the DNA double helix displayed above the sketch of a human body and the depiction of solar planetary system at the bottom, with largest square representing Sun and small outlier square representing Earth—the home planet of the message senders

message, and read its contents, thus proving that the DNA "message" safely survived during its journey via the common mail, and therefore successfully verifying the new encryption technique they invented.

The ability of DNA to encode messages was also noted by those scientists who, based on certain evidences, considered the possibility that terrestrial life is not originated on our planet but has been seeded intentionally by some extraterrestrial civilization from the faraway galaxy. They reasonably argued that if this would indeed be the case, then, without any doubts, the DNA of some simple terrestrial organisms, like bacterial viruses (aka bacteriophages) or spore-forming bacteria—the most probable candidates for safe interstellar travel[1]—must carry the coded message from an advanced society, which sends them through the space [4, 5]. Such a message could be in the form of encoded short text embedded into particular DNA sequence, or it could be an encrypted two-dimensional image (aka pictogram or pictograph) with visual information about the sender's society that can be drawn from that sequence when decoded [4]. Most probably, the alien pictogram would display readily recognizable hallmarks of artificiality, i.e., some simple signs, like cross (X), square, or triangle, or the symbol of zero.

Several attempts have been undertaken to find an encoded message from an extraterrestrial intelligence in DNA of terrestrial microorganisms. For instance, Japanese scientists considered bacteriophage ϕX174 DNA as a candidate interstellar message in a form of a pictogram. Accordingly, they tried to decode such a

---

[1] Experiments proved that bacteriophages and bacterial spores, which are extraordinarily resistant living species, can survive in the harsh environment of outer space if they are laid within meteorite rocks to be shielded from harmful cosmic radiation [6]. These rocks would also protect microorganisms from intense frictional heat during a high-speed passage through the Earth's atmosphere [7].

pictogram from genetic sequences of certain φX174 proteins [4]. Unfortunately, no significant simple patterns were decrypted from these sequences (see Fig. 2.3a for one of the pictograms obtained). And even if we are willing to assume that this is some data matrix code sent to us by aliens, no reasonable text message can be deciphered from it, alas.

It is worth mentioning that experimental attempt in contacting potential alien civilizations via pictograms was undertaken by American astronomers in 1974 when the pictographic image carrying basic information about humankind and Earth was encoded in powerful radio signal to be sent from the world's largest radio telescope in Arecibo, Puerto Rico, toward globular star cluster M13 (aka the Great Hercules Cluster). M13 is located 25,000 light years away, being one of the few closest-to-Sun large collections of hundreds of thousands of stars (see Fig. 2.3b), and it is assumed to be a likely place for alien civilization to inhabit some planet which might exist there. The interstellar radio message sent from Arecibo encrypted a 2D diagram showing, among other things, a human body, our solar system, and the DNA molecule—the essence of terrestrial life (Fig. 2.3c)—in the hope that extraterrestrial intelligence might receive and decipher it.

Though DNA is thought as a perfect messenger for intergalactic communications, along with radio transmissions [8, 9], so far no traces of any kind of intelligent signatures in DNA sequence of any creature living on Earth have been found, thus leaving the intriguing question "Are we from Earth or from outer space?" as open.

To conclude this section, I would like to note that hypothetical alien's messaging with the use of microbial DNA suggested by Japanese scientists 40 years ago [4] is now not a pure theory: in 2004 two other Japanese scientists encoded the word "KEIO" (the institution name of the authors' affiliation) within synthetic protein-coding sequence[2] and introduced it in the genome of spore-forming bacteria *Bacillus subtilis* [10]. The "signed" (or "watermarked") bacteria grew and sporulated normally, and in its spore form it is resistant to extremes of environment including perhaps those of interstellar travel.

As to more earthy goals, the innovative technology of encrypting genomic DNA can be used to help establish brand names for the strains of all commercialized engineered bacteria [10, 11] and also to resolve legal disputes regarding gene-related patents. And this is only one example of different ways and uses of emerging techniques of coding objects with DNA, which will be the topic of next section.

2.2 DNA Identity Tags

Another possible use of information that can be encoded in synthetic DNAs is to unnoticeably mark various objects with specific nucleotide sequence (or set of specific sequences) for identification/authentication purposes [12]. The imaginary

---

[2] Instead of encryption code, digital binary encoding scheme was used in this experiment, which will be discussed below in Sect. 2.3.

realization of this idea was depicted in sci-fi movies *Judge Dredd* (1995) and *Dredd* (2012) showing a fictional weapon Lawgiver used by the Judges.

The Lawgiver is a handgun (Fig. 2.4), which fires bullets labeled with the DNA code of the Judge to whom the weapon belongs, thus making easy identification of the shooter. The Lawgiver II, a revised and more personalized version of this handgun, is also tagged to the DNA of its designated owner. If someone other than an authorized judge attempts to fire the gun, it self-destructs by detonation injuring an unauthorized user. Inspired by these Hollywood's fictional weapons, Australian firearm company DefendTex and New Jersey Institute of Technology determined to turn science fiction into reality: by using their proprietary biometric authentication methods, they are both developing smart guns designed to be used only by its designated owner, thus ensuring that it cannot be fired if lost or stolen.

Several other companies are developing DNA labels to mark a variety of civilian objects to prevent or detect their theft or counterfeit. Examples include the DNA marking of banknotes, credit cards, precious jewelry, works of art, and other articles having financial value. Also, the DNA labels on computers and clothes would help to undoubtedly identify the country of origin where these products were produced or assembled to control their legal vs. illegal import. They would also help to identify outerwear and footwear faking popular brands of clothes and shoes. Encoding, detection, and verification of DNA labels can be done similar to described above for DNA-encoded secret messages.

Art experts estimate that over half of all art pieces that are on the market today are actually fake. To fight art forgers, the British company Tagsmart has recently pioneered the application of synthetic DNA tags onto canvasses and art prints to verify and certify their authenticity. Tamper-proof Smart Tags, as they are called, stick tiny amounts of synthetic DNA of unique sequence to the artworks with an adhesive beneath a 1.5-in.-diameter flexible tag that is also packed with several high-tech security features to be easily noticed if it is tampered with. The tags are affordable: prices range from $25 for art-print tag to about $150 canvas tag, and many artists started tagging their artworks with Smart Tags so that collectors would know with absolute certainty that the painting they purchase is authentic.

Fig. 2.4 Justice Dept. Lawgiver MK II. (Computer-generated image by Mike Carroll reproduced under GNU Free Documentation License from https://en. wikipedia.org/wiki/ File:Lawgiver-mkii.jpg)

The business of counterfeit goods is one of the largest underground businesses in the world, and it is growing rapidly having the negative and harmful effects both on industries and on consumers. In 2008, the US Government estimated the global market value of the counterfeit industry at $500 billion with a growth rate of 1.7% over the past 10 years. As an example, a few years ago the German Customs Department seized what could have been the largest cache of counterfeit goods—1 million pairs of phony Nike sneakers, in a total of 117 shipping containers worth nearly $500 million!

To combat this large problem, Applied DNA Sciences Inc., an American high-technology company based in Stony Brook, NY, is developing the DNA textile markers to label both textile raw materials and finished goods [13]. These markers can be used to verify authenticity and to protect trademarked textile products and related intellectual property that are critical to the textile product industry. Importantly, DNA markers are visibly undetectable (as only trace amounts of DNA are required to be spotted onto some place for reliable retrieval from it by PCR) and it is impossible to fake them unless you know the secretive sequence of the marker. The long-term stability of DNA chemical structure at normal and elevated temperatures proved by several studies [14–16] would also ensure the preservation of DNA labels in their intact form for some considerable time (possibly years) even under uncontrolled storage conditions. The company therefore believes that its DNA markers will likely be able to withstand, especially if additionally protected from any potential environmental degradation [13], extremely harsh textile processes, such as bleaching, mercerizing, dyeing, and finishing, as well as hot and wet weather, and will remain embedded in the fabric or yarn for years [17, 18].

Furthermore, synthetic DNAs can be used as a convenient and cost-effective tracer to assess the release of pollutants into groundwater or other aquatic resources [19]. BaseTrace, the North Carolina-based biotech startup, is developing DNA tracer technology to track possible contamination of natural waters from hydraulic fracturing (aka fracking). Fracking is a drilling process widely used in the United States and other oil- and gas-producing countries, where millions of gallons of water containing highly reactive and hazard chemicals are pumped underground to break apart the rock or hard soil and to release the crude oil or gas. While this technology is very efficient in the extraction of these valuable natural resources, it has become a contentious environmental and health issue because of enormous potential risks of contamination of drinking water sources near drilling sites due to possible leaks from oil or gas extraction wells to underground water used as a drinking water supply. Likewise, nuclear power plants are also concerned about leak detection.

Many dyes and other existing tracers are either costly or chemically incompatible with drilling and nuclear conditions, so detecting a leak may be problematic in most cases. In contrast, the BaseTrace tracers are composed of inexpensive encoded pieces of DNA that can be mixed with a wide variety of industrial fluids, like hydraulic fracturing fluid or nuclear reactor coolant, providing each fluid source with a chemical fingerprint that is simple and cheap to identify by PCR amplification and sequencing [20]. This would enable to readily find leaks and trace them back to their original fluid source.

Returning to DNA markers, I have to say that besides labeling objects with specific DNA sequences there is one more option for DNA encoding called DNA barcoding or digital DNA coding [21]. Similar to ordinary barcoding that uses unique patterns of printed parallel lines of varying widths and positions to distinctively mark various objects (as it is shown in Fig. 2.5a), the object of interest can be tagged with a set of DNA fragments of different lengths flanked by common priming sites for PCR amplification.

To expose and to read the barcode, DNA fragments are eluted from the labeled spot on the object and amplified by the PCR with a universal pair of primers. Then, the amplified fragments are resolved by gel electrophoresis (see Glossary for explanation of this DNA analysis technique) and the gel is optically scanned to get the object's barcode represented by thus obtained set of bands, as it is schematically shown in Fig. 2.5b.

More than a million different DNA barcodes can be created this way to make the method practically convenient for unique labeling of multiple objects. And the reliability of this barcoding scheme has been demonstrated by its inventor, British researcher Jonathan Cox, in a similar way as described above for DNA-encoded messages by labeling a letter with a minute amount of DNA fragment mixture, sending the letter through the mail and successfully reading the barcode on receipt of the letter [21].

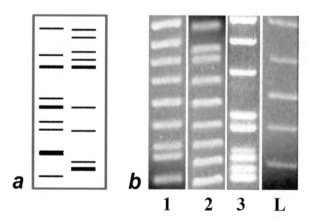

Fig. 2.5 (**a**) Two exemplary conventional nine-band printed barcodes differing by spacings and widths of parallel lines. (**b**) Examples of the three different nine-band DNA barcodes visualized by gel electrophoresis (lanes 1–3); lane L is the DNA ladder serving as a barcode ruler. Note that the width of DNA bands can also vary by using the two or more close-length DNA fragments to act as the thick elements of conventional barcodes (like the two almost unresolved bands at the bottom of lane 3). Additionally, brightness of bands can vary, too, by loading different amounts of DNA fragments, as it can be seen for the faint fourth fragment from the bottom of lane 3, thus easily providing extra complexity to DNA barcoding

2.3   The Rise of DNA Data Storage

Why cannot we write the entire 24 volumes of the Encyclopaedia
Brittanica on the head of a pin?
Richard P. Feynman (American physicist), There's Plenty of Room
at the Bottom: An Invitation to Enter a New Field of Physics *(1959–1960)*

Besides the coding in DNA sequence of simple texts and barcoding with DNA fragments, DNA can be employed as a medium for digital binary encoding, which is commonly used in electronic computing and digital communication. In this way, essentially all types of information, including visual and acoustic information, can be encoded in a sequence of two digits "0" and "1," which in turn will be converted into encoded sequence of four DNA bases.[3]

In the simplest encoding scheme, each of the DNA bases can be encoded by the two binary digits (i.e., 2 bits), like A = 00, T = 01, G = 10, and C = 11. This would allow encoding in DNA vast amounts of information: 2 bits of information per nucleotide means that it is theoretically possible to embed as much of information into just 1 g of DNA as about 4 sextillion bits (i.e., 4×10^{21} bits or 4 zettabits; for the names of large numbers, which you will encounter in this chapter, see Table 2.2). To better understand how big is this number it is worth to compare it with a similar amount of all the digital data transferred across the globe daily! Or just consider that such a little piece of "organic" data memory is capable of storing about the same amount of data as trillion (10^{12}) CD-ROMs. By other estimates, a few kilograms of DNA could theoretically store all of the world's data. And to read the data stored in these informational DNAs (iDNAs), you have to simply sequence them and to convert each of the A, T, G, and C bases back into binary digits.

The very first proof that extrabiological information in the form of graphical symbol can be written into DNA via binary code has been made in the late 1980s by artist and researcher Joe Davis (who described all details of this experiment later in his 1996 article [23]) with the help of his collaborator, molecular geneticist Dana Boyd at Harvard Medical School (Boston, USA). Davis decided to encode into

Table 2.2 Names of large numbers

Name	Value	Prefix (Unit symbol)
Million	10^{6}	mega-(M)
Billion	10^{9}	giga-(G)
Trillion	10^{12}	tera-(T)
Quadrillion	10^{15}	peta-(P)
Quintillion	10^{18}	exa-(E)
Sextillion	10^{21}	zetta-(Z)

[3] The general idea of using DNA as a minute information storage tool was originally proposed in mid-1960s by American mathematician Norbert Wiener, one of the founding fathers of cybernetics, and independently by Russian physicist Mikhail Neiman, who also contemplated at the same time the possibility of storing and retrieving data in DNA molecules [22].

bacterial DNA a graphic symbol for life and femininity, which can be represented by a letter Y from ancient German alphabet (aka runic alphabet) symbolizing "life/fertility" and called the "life rune," and can also be seen in ancient Mycenaean psi(ψ)-shaped female figurines (Fig. 2.6a).

This two-dimensional female gender symbol (called *Microvenus* by Davis) was first translated into a look-alike five-by-seven table of "0" and "1" digits (see Fig. 2.6b). Davis cleverly noted that the number 35, the product of 5 and 7 (i.e., numbers of columns and rows in the table), is divisible only by those two prime numbers. Therefore, he assumed that the meaning of the 35-digit binary number 10101 01110 00100 00100 00100 00100 00100 generated for insertion into DNA and obtained by the reverse transformation of Y-like digital table into linear sequence of 0s and 1s would be readily recognized by some hypothetical decoder out there in the universe (note here that Y also symbolizes antenna in electronic diagrams). So it can be converted back to only two 2D numerical tables—one is the five columns-by-seven rows symmetrical Microvenus-like table shown in Fig. 2.6b, and another one would be the seven columns-by-five rows asymmetrical "nonsensical" table

1010101
1100010
0001000
0100001
0000100

which does not resemble any reasonable figure or symbol.

Using the phase-change coding method (a bit more complicated code than one described above), the 35-digit Microvenus-encoding binary number was further translated to only 18 DNA bases, CCCCCCAACGCGCGCGCT, and appended at the beginning by the decoding clue, CTTAAAGGGG, for this particular code [23]. The combined 28-base-long DNA sequence, CTTAAAGGGGCCCCCCAACGCGCGCGCT, which

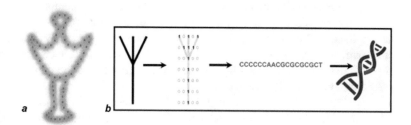

Fig. 2.6 (**a**) Typical outline of a psi(ψ)-shaped figurine of a woman. This type of figurines has been made of molded and painted terracotta in ancient Greece circa thirteenth century BCE and they were found in tombs, shrines, and settlement areas [24, 25]. Possible uses were children's toys, votive figurines, or grave offerings. (**b**) Transformation of the ψ-like graphic symbol of female gender into sequence of DNA nucleobases for insertion into double-helical bacterial genomic DNA (red stripes symbolize the inserted sequence and its complement). Arrangement of 1s in this binary digit table resembles ψ shown as a light-colored background (see text for details)

carries both Microvenus-encoding "word" and clue for a letter-to-digit decoding, was inserted by Dana Boyd into circular DNA vectors called plasmids using well-known biotechnological methods. And these vectors were finally cloned into several laboratory strains of *E. coli* bacteria.

In subsequent "post-Microvenus" years, several other proof-of-principle studies on digital information storage in iDNAs have been published that were based on different data coding schemes and different data storage architectures [10, 11, 26, 27], but they all stored only small amounts of data (reviewed in [28]).[4] Then in 2012, the breakthrough study has been published by the trio of scientists from Harvard Medical School and Johns Hopkins University (Baltimore, USA) reporting an efficient approach for encoding in iDNAs a vast amount of diverse digital information employing advanced DNA synthesis and sequencing technologies [30]. Primary advantages of the new "data-storage-in-DNA" approach over the prior approaches include one-bit-per-base encoding scheme (G or T as "1" and A or C as "0") and conversion of data into a sequence of bits (known as a bitstream), splitting the stream of bits into data blocks and storing blocks of digital data in multiple copies of short DNA pieces, called oligonucleotides, ordered and arrayed on the surface of glass slides (aka DNA microarrays or DNA microchips).

Though one-bit-per-base encoding is twice less dense than the two-bit-per-base encoding described above, this alternative coding scheme, along with avoiding the use of long iDNAs, circumvents certain DNA sequences which are challenging to be written or read. As a result, this inventive DNA encoding strategy made it practically possible to convert the whole textbook composed of more than 50,000 words, and also containing 11 images and a JavaScript program, into the ordered set of 55,000 oligonucleotides spotted on DNA microchips [30]. To read the book encoded in iDNA, scientists decoded it with just a few errors by using high-throughput next-generation sequencing technology developed by biotech company Illumina, Inc.

In this seminal experiment, more than 5 million bits (or megabits) of information was encoded in DNA at the very high density of 5.5×10^{15} bits (5.5 petabits) per mm^3 of DNA microchips or 10 petabits per gram of DNA spotted on microchips. This is million times denser than it could be achieved in the flash memory devices, which currently provide us with the greatest density of stored information in commercial appliances. Nevertheless, the encoding scheme employed by Harvard-Hopkins team, though much more robust than those used in all such previous studies, was relatively inefficient since the achieved high information density was still so far from theoretically possible one.

Recently, a new study on data storage in DNA has been published by the two researches from New York Genome Center (NYGC), which exploited different and more efficient data encoding scheme and data storage architecture [31]. This advanced DNA encoding strategy, named DNA Fountain, is based on the use of so-called fountain codes developed for reliable and effective delivering of mobile TV

---

[4] In 2002, the group of researchers from Duke University (Durham, USA) intended to build very large, petabit-sized ($\sim 10^{15}$ bits) DNA-based data storage [29], but this promising project has never been completed.

services. Basically, it tightly packages digital data into a number of short dense messages, called binary droplets. Then, the specially designed algorithm translates these digital droplets to DNA sequences by converting {00,01,10,11} to {A,C,G,T}, respectively, to create a pool of short DNA segments (oligonucleotides), which thus encode the incoming digital data. Consequently, by using a standard laptop and DNA Fountain encoding, it took NYGC researches only 2.5 min to encode a full computer operating system, movie, and other files with a total of 10 megabits of information for storing all this information in the pool of 72,000 DNA oligonucleotides and then perfectly retrieved the information by Illumina sequencing.

The new method of storing data in DNA is the highest-density storage scheme ever invented, as it was proved to be capable of storing near 2 exabits in a single gram of DNA. Such a density is over two orders of magnitude higher than reported by Harvard-Hopkins team with a comparable number of oligonucleotides and it is closer to the estimated theoretical upper limit. Thus, given that average book contains about 10 megabits of information, the contents of the whole library as big as British library that holds around 170 million items can now be encoded and stored in a minute-size library of DNA oligonucleotides weighting only 1 mg (a grain of sand!). Remarkably, this likely achievement will overwhelmingly overcome the clever Feynman's idea of writing the contents of two dozen books on the head of a pin by photoelectric engraving [32][5] (see his quotation at the beginning of this section). Indeed, the microscopic engraving of entire British library would need, by Feynman's own estimates, more than a million pinheads arranged in the area of several square yards, which is in drastic contrast to 1 mg of DNA required for the storage of the same amounts of data.

Yet, we should not expect the replacement of CDs or DVDs with video or audio iDNA chips anytime soon: the maximal current speed of DNA sequencing (which in third-generation DNA sequencers could be close to 250 nucleotides per second, or 1000 bits per second in informational terms) is very far from the currently achievable reading speed of CD/DVD drives (up to 1000 megabits per second).[6] Still, iDNAs would be an excellent material for very-long-term storage of huge volumes of data. As a matter of fact, the survival of ancient bacterial spores over many millions of years, as they were revived undamaged from very old, prehistoric inclusions in buried salt crystals and amber [33, 34], validates the superior durability of genetic information preserved in bacterial DNAs, as compared to rather short-term durability of data stored on various magnetic or optical disks which, by different estimates, have a life span of 20–100 years only. And the recent experiment has directly demonstrated that glass-encapsulated iDNA can indeed serve as the ultimate time capsule: a team of Swiss researchers were able to error-free retrieve all digital data they translated to the silica-entrapped DNA oligonucleotides after keeping them at 70 °C for 1 week, which is equivalent to 2 million years when compared to storage conditions at −18°C [35].

---

[5] In this paper, renowned physicist Richard Feynman outlined the concepts of nanotechnology.

[6] It is however possible that in the future certain innovative technologies would make it possible the faster reading of DNA sequences similar to the reading of magnetic tapes.

I should note here that a few years ago the Japanese company Hitachi disclosed that their developing proprietary technology for storage of information in small pieces of quartz glass could also keep data safe for million years. However, the amount of data a new tool might be able to hold is not as large as iDNA chips can store: this alternative data-recording medium has a storage density only slightly better than that of a CD [36]. The superiority of DNA for storing digital data has recently attracted attention of major tech companies, like Amazon, Google, Facebook, Apple, and Microsoft, which are now making their initial investments into the commercialization of this cutting-edge promising technology. For instance, the global software giant Microsoft is developing proto-commercial DNA-based storage device with the intended size of a commercial Xerox copier for pilot exploitation in one of its data centers [37].

Recently, Microsoft researchers in collaboration with scientists from the University of Washington (Seattle, WA) have tested one of the possible solutions towards this goal [38]. They stored dehydrated iDNA with data encoded into nucleotide sequences as multiple isolated and densely arranged micro-spots on small glass plates, which could be further organized into a rack of decks to enable high information storage density. Furthermore, to retrieve data they constructed portable digital microfluidic (DMF) cartridge that is only a few inches in size, and which manipulates the fluids as droplets by using electrical voltage and an array of electrodes [38, 39]. For data retrieval, the DNA-spotted glass plate is loaded onto the DMF device controlled by a computer to move an aqueous droplet to specific location on a plate for liquifying particular DNA spot and then to move droplet with the dissolved DNA to a DNA sequencer (Fig. 2.7).

It was demonstrated in this study that more than a terabit of data could be stored in a single tiny spot of DNA and successfully retrieved by using the DMF technology. It is feasible that several hundred micro-spots can be orderly deposited per glass plate and several hundred thin plates can be readily assembled in a rather compact rack, thus leading to an exabit storage capacity of such a device. Individual plates could be taken out from the rack of decks by a robotic device, similar to how tape and disk drives are stored and handled in data centers today. Altogether, this kind of system could realize the oft-hyped, much-demanded data density of DNA-based data storage.

Of course, not all innovative technologies end up becoming widely used beyond just research programs. But without any doubts our modern magneto-electro-optical data storage tools will soon become outdated due to low data density of disk-, stick-, or tape-based systems and their relatively short lifetime. On the other hand, future breakthroughs in DNA synthesis and DNA sequencing that are expected within the next decade could make it possible for us to buy and to use an inexpensive DNA memory card instead of disk or flash drives (Fig. 2.8).

Moreover, encouraged by breakthrough developments in DNA data storage technologies material science researchers are extending molecular-scale information-coding studies on macromolecules other than DNA in a hope to find even better solutions for storing the data. But such emerging technologies are beyond the scope of this book so I just refer the interested reader to one of the latest proof-of-principle

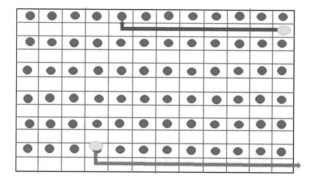

Fig. 2.7 Schematics of the DNA data storage system based on DMF technology. A rectangular glass plate is regularly spotted with DNA micro-spots (small blue circles), each encoding a piece of particular digital data and having a specific "address" on a plate. To retrieve the DNA-encoded data, this glass plate is loaded onto microfluidic cartridge comprising an array of electrodes (shown as small rectangles visible through glass). Such an array is digitally programmed by a computer to move, by using electrical voltage, aqueous droplets (grey ovals) sandwiched between the cartridge and glass plate from one electrode to another, and finally to a specific location on a plate (shown by red arrow). Then, after dissolving desired DNA micro-spot(s), droplets are moved out of a plate (shown by green arrow) to a DNA sequencer to read DNA sequences and thus decode the data

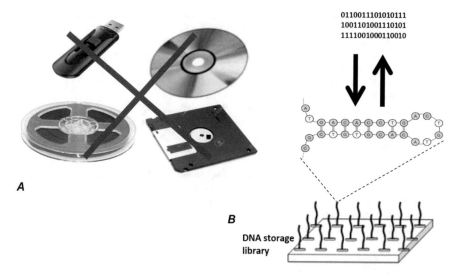

Fig. 2.8 The future of digital data storage? It may sound futuristic, but given that the key measures of digital and biological technologies are improving at exponential rates related to Moore's law, including the processing speed, size, cost, and density of components [40, 41], DNA could be the ultimate eternal memory card of the future. It is quite imaginable that, once becoming obsolete, common magnetic, electrical, and optical storage devices shown in (**a**) will be replaced someday by DNA chips as shown in (**b**) encoding and storing digital information in DNA sequences. Depicted here is hairpin-DNA memory that enables the temperature-controlled repetitive data writing and erasing [42]

studies, which demonstrates that digital information can be stored as a sequence of molecular units (monomers) in macromolecules other than DNA and that thus encoded data may be read by conventional fast-response analytical instrumentation, such as mass spectrometry [43].

Also of note here is that serving as an information storage medium is not the only DNA ability for handling and processing digital data. In the next chapter, it will be shown that based on the DNA's ability to build nanostructures with molecular precision DNA can be employed in binary and logical computing, too, instead of the traditional silicon-based computing technologies. So, if you are inclined to continue reading, I hope, do not miss next the piece about this intriguing subject!

References

1. Hayden EC (2015) Pint-sized DNA sequencer impresses first users: portable device offers on-the-spot data to fight disease, catalogue species and more. Nature 521:15–16
2. Nanopore (2019) MinION Brochure. https://nanoporetech.com/sites/default/files/s3/literature/MinION-Brochure-14Mar2019.pdf
3. Clelland CT et al (1999) Hiding messages in DNA microdots. Nature 399:533–534
4. Yokoo H, Oshima T (1979) Is bacteriophage ϕX174 DNA a message from an extraterrestrial intelligence? Icarus 38:148–153
5. Hoch JA, Losick R (1997) Panspermia, spores and the *Bacillus subtilis* genome. Nature 390:237–238
6. Olsson-Francis K, Cockell CS (2010) Experimental methods for studying microbial survival in extraterrestrial environments. J Microbiol Methods 80:1–13
7. Fajardo-Cavazos P, Link L, Melosh HJ, Nicholson WL (2005) *Bacillus subtilis* spores on artificial meteorites survive hypervelocity atmospheric entry: implications for lithopanspermia. Astrobiology 5:726–736
8. Shcherbak VI, Makukov MA (2013) The "Wow! signal" of the terrestrial genetic code. Icarus 224:228–242
9. Marx G (1979) Message through time. Acta Astronaut 6:221–225
10. Arita M, Ohashi Y (2004) Secret signatures inside genomic DNA. Biotechnol Prog 20:1605–1607
11. Gibson DG et al (2010) Creation of a bacterial cell controlled by a chemically synthesized genome. Science 329:52–56
12. Bancroft C, Clelland CT (2001) DNA-based steganography. US patent 6,312,911
13. Jung L, Hayward JA, Liang MB (2014) DNA marking of previously undistinguished items for traceability. US patent application US20140272097A1
14. Radzicka A, Wolfenden R (1995) A proficient enzyme. Science 267:90–93
15. Karni M et al (2013) Thermal degradation of DNA. DNA Cell Biol 32:298–301
16. Integrated DNA Technologies (2014) Oligonucleotide stability study. https://sfvideo.blob.core.windows.net/sitefinity/docs/default-source/technical-report/stability-of-oligos.pdf?sfvrsn=c6483407_10
17. Paunescu D, Fuhrer R, Grass RN (2013) Protection and deprotection of DNA—high-temperature stability of nucleic acid barcodes for polymer labeling. Angew Chem Int Ed Engl 52:4269–4272
18. Fabre AL et al (2017) High DNA stability in white blood cells and buffy coat lysates stored at ambient temperature under anoxic and anhydrous atmosphere. PLoS One 12:e0188547
19. Sabir IH, Torgersen J, Haldorsen S, Alestrom P (1999) DNA tracers with information capacity and high detection sensitivity tested in groundwater studies. Hydrogeol J 7:264–272

20. Chow JS, Rudulph JG (2014) Systems, methods, and a kit for determining the presence of fluids of interest. International patent application WO2014005031A1
21. Cox JPL (2001) Bar coding objects with DNA. Analyst 126:545–547
22. Neiman MS (1965) On the molecular memory systems and the directed mutations. Radiotekhnika (Russian) 20(6):1–8
23. Davis J (1996) Microvenus. Art J 55(1):70–74
24. French E (1971) The development of Mycenaean terracotta figurines. Annu. Br. Sch. Athens 66:101–187
25. Olsen BA (1998) Women, children and the family in the late Aegean Bronze Age: differences in Minoan and Mycenaean constructions of gender. World Archaeol 29:380–392
26. Skinner GM, Visscher K, Mansuripur M (2007) Biocompatible writing of data into DNA. J Bionanosci 1:1–5
27. Ailenberg M, Rotstein OD (2009) An improved Huffman coding method for archiving text, images, and music characters in DNA. BioTechniques 47:747–754
28. De Silva PY, Ganegoda GU (2016) New trends of digital data storage in DNA. Biomed Res Int 2016:8072463
29. Reif JH et al (2002) Experimental construction of very large scale DNA databases with associative search capability. Lect Notes Comput Sci 2340:231–247
30. Church GM, Gao Y, Kosuri S (2012) Next-generation digital information storage in DNA. Science 337:1628
31. Erlich Y, Zielinski D (2017) DNA Fountain enables a robust and efficient storage architecture. Science 355:950–954
32. Feynman RP (1960) There's plenty of room at the bottom: an invitation to enter a new field of physics. Eng Sci 23(5):22–36
33. Cano RJ, Borucki M (1995) Revival and identification of bacterial spores in 25 to 40 million year old Dominican amber. Science 268:1060–1064
34. Vreeland RH, Rosenzweig WD, Powers DW (2000) Isolation of a 250 million-year-old halotolerant bacterium from a primary salt crystal. Nature 407:897–900
35. Grass RN et al (2015) Robust chemical preservation of digital information on DNA in silica with error-correcting codes. Angew Chem Int Ed Engl 54:2552–2555
36. Hornyak T (2013) Super long-term storage. Sci Am 308:21
37. Regalado A (2017) Microsoft has a plan to add DNA data storage to its cloud. MIT Technology Review 120. www.technologyreview.com/s/607880/microsoft-has-a-plan-to-add-dna-data-storage-to-its-cloud
38. Newman S et al (2019) High density DNA data storage library via dehydration with digital microfluidic retrieval. Nat Commun 10:1706
39. Alistar M, Gaudenz U (2017) OpenDrop: an integrated do-it-yourself platform for personal use of biochips. Bioengineering 4:45
40. Moore GE (2006) Progress in digital integrated electronics. IEEE Solid-State Circuits Society Newsl 9:36–37
41. Carlson R (2003) The pace and proliferation of biological technologies. Biosecur Bioterror 1:203–214
42. Takinoue M, Suyama A (2006) Hairpin-DNA memory using molecular addressing. Small 2:1244–1247
43. Al Ouahabi A, Amalian JA, Charles L, Lutz JF (2017) Mass spectrometry sequencing of long digital polymers facilitated by programmed inter-byte fragmentation. Nat Commun 8:967

DNA as a Nanoscale Building Material

3

The Watson-Crick complementarity of DNA strands that leads to their pairing into the DNA double helix, and that enables DNA to serve so effectively as genetic material, can be used for other purposes, too. This chapter describes the awesome skill of DNA to direct the assemblies of more complex DNA structural motifs than simple double, triple, or quadruple helical forms of DNA described in the introductory chapter of this book.

Using this mighty DNA ability, researchers have now created a menagerie of diverse nano-sized DNA structures. And it all started in early 1980s, when two American biochemists Nadrian Seeman and Neville Kallenbach have realized that synthetic DNA sequences can be designed to combine and to fold them in certain ways, and thus to be programmed for highly specific routes to building DNA nanostructures [1, 2]. This seminal perception gave the birth to structural DNA nanotechnology—manufacture of artificial extremely small, i.e., nano-sized nucleic acid structures for technological uses, with applications in biosensing, drug delivery, biomolecular analysis, and molecular computation, to name just a few [3].

3.1 Building Nano-Objects with DNA Construction Sets

I believe that everybody enjoyed in the childhood by playing with toys called construction sets—collections of standardized pieces that allow for the construction of a variety of different objects, which are typically a few inches long (Fig. 3.1). Amazingly, you can do this with DNA on a molecular scale. You only need a set of certain DNA-building modules (or building blocks), and optionally a protein, called DNA ligase enzyme, to covalently link them into a desired shape.

The very first designed DNA construction sets include two principal structural motifs (Fig. 3.2). As building blocks, these sets contain branched DNA junction complexes with three and/or four arms (aka three- and four-way junctions), which can be readily assembled from sufficiently long stretches of synthetic DNAs by base-pairing. Importantly, these primary complexes should have sticky ends, which

Fig. 3.1 Examples of popular construction set toys. Interlocking plastic Lego bricks (**a**) and magnetic Goobi bars and balls (**b**); some simple constructs assembled from them are shown on the right

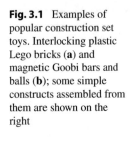

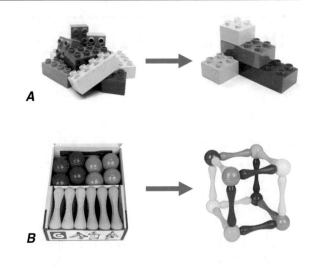

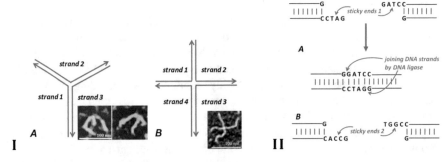

Fig. 3.2 Two principal structural motifs of DNA construction sets. (**I**) Schematic representation and micrographs of branched DNA junctions with three (*A*) and four arms (*B*). They are formed by complementary base-pairing of three and four DNA strands arranged in antiparallel orientation to each other (shown by arrows). When interconnected, these junctions could serve as vertices (i.e., the corner parts) of various geometric shapes, such as tetrahedron, octahedron, and gridiron, since they are flexible enough to form specific angles at their branches required by spatial geometry of specific objects [4–6] (see insert in schematics **I***A* showing the atomic force microscopy images of the two three-arm DNA junctions with apparent tetrahedral shapes). Currently assembled DNA branched junctions can have up to 12 arms [7, 8] so that essentially unlimited number of various objects, graphs, and networks can be built from branched DNA components. (**II**) Sticky DNA ends act as highly selective molecular glue. (*A*) Pieces of duplex DNA can be joined together by complementary base-pairing of compatible single-stranded overhangs (called cohesive ends or sticky ends) working as a Scotch tape, followed by creation of phosphodiester bonds at the two single-stranded breaks (aka nicks) in thus formed extended DNA duplex by enzyme DNA ligase. (*B*) A different pair of compatible sticky DNA ends, which can only join the corresponding duplex DNA fragments, but not DNA fragments with dissimilar overhangs. (Microscopy images in **I***A, B* are the courtesy of Yurii Lyubchenko, University of Nebraska)

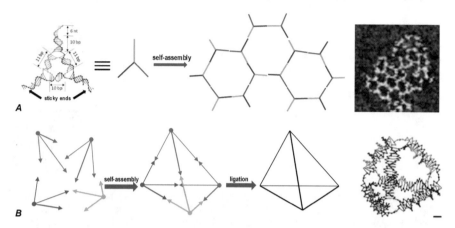

Fig. 3.3 2D (**a**) and 3D (**b**) nanoscale DNA assemblies resulted from three-arm DNA junctions. (**a**) The plurality of three-arm junctions with sticky ends formed by hybridization of three linear and one circular oligonucleotides, as it is shown on the left, can build the honeycomb-like mesh shown schematically in the center. Using the pre-circularized oligonucleotide as the core strand of DNA junctions makes them less flexible. Thus, it provides DNA junctions with sufficient rigidity to stay predominantly in a planar state, which is a prerequisite for growing protracted 2D DNA lattices. The micrograph of thus created molecular mesh with the unit cell size of about 15 nm is shown on the right. (Adapted from [9].) (**b**) Schematics of possible assembly of a transparent DNA tetrahedron (aka triangular pyramid) from the four flexible three-arm junctions with sticky ends (for simplicity, single lines represent double-stranded DNA arms and the arrows represent sticky ends). It is known that flexible DNA junctions are very dynamic structures that can form interarm angles as low as 60° and less [4–6], which is necessary for them to be connected into a tetrahedral shaped object. If the length of junction's arms is ~15–20 bp as above, the height of such a tetrahedron would be ~10 nm. A model of DNA tetrahedron with the three-arm junctions at each vertex, which could be obtained this way, is shown on the right (*made with NanoEngineer-1 software by John P. Sadowski and reproduced under free license from https://en.wikipedia.org/wiki/File:DNA_tetrahedron_white.png*). In this model each base pair is represented by five pseudo-atoms, representing the two sugars, the two phosphates, and the major groove of the DNA double helix. The scale bar at bottom right is 1 nm

will act as a molecular glue to link the basic set of DNA branches in a prearranged way, thus forming more complex 2D and 3D assemblies.

Accordingly, by having such a DNA construction set one can assemble a variety of different 2D or 3D nanoscale objects, networks, and complex patterns out of DNA [7–12].[1] The instruction to do so is rather simple: take synthetic multi-arm branched DNA molecules with programmed sticky ends, and get them to self-assemble into the desired structure, which may be a 2D array or a 3D object (Figs. 3.3 and 3.4a). Generally, the sticky end-directed DNA assembly allows multiple DNA parts to be assembled together in one reaction container from a single mixture of necessary components (so-called one-pot reaction or one-pot synthesis).

---

[1] In particular, three- and four-way junctions were used in early 1990s by Nadrian Seeman and his coworkers at New York University (NYU) in the pioneering studies leading to constructions of a multistranded DNA cube and truncated DNA octahedron, the two very first 3D nanoscale objects [11, 12].

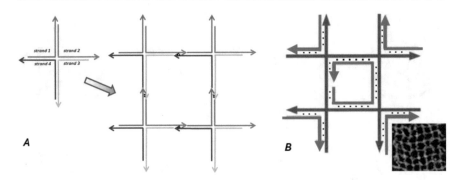

Fig. 3.4 Assemblies of two-dimensional gridiron-like DNA nanostructures based on four-arm DNA junctions. (**a**) Schematic representation of four DNA strands annealed to form a DNA branched junction with sticky ends, which allows further self-assembly of these junctions into a nanoscale square mesh (or nanogrid). This diagram shows geometry and strand polarity of a single gridiron unit formed from four 4-arm junctions. (**b**) Alternative design for assembly of DNA grid-iron based on four-arm junctions formed with the use of short single-stranded DNA loops. Insert is the micrograph of thus created molecular mesh with 20 × 20 nm square units. (Courtesy of Dongran Han, Arizona State University)

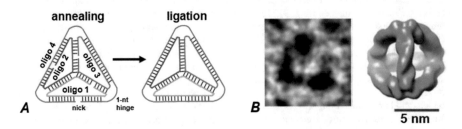

Fig. 3.5 Building the DNA tetrahedron from a set of specially designed oligonucleotides. (**a**) The annealing of four oligonucleotides results in their self-assembly into tetrahedral structure. (**b**) The left image is an original micrograph of thus formed DNA tetrahedron; shown on the right is the 3D density map of this DNA construct obtained by software analysis of similar microscopy images, which explicitly reveals the tetrahedral structure with double-helical edges. (Adapted from [14])

You may also add a DNA ligase to make thus self-assembled structure a single giant DNA molecule with the covalently closed shape you wanted. Another approach for assembly of DNA nanostructures based on multi-arm junctions is shown in Fig. 3.4b and it would require longer DNA strands but less ligation points [10].

Inspired by the success of original modular strategy based on multi-arm DNA junctions that resulted in the assembly of a variety of discrete DNA nanostructures and nanoscale networks (some of which are presented above), researchers developed a number of alternative DNA building blocks that significantly increase the range of DNA-based constructs and simplify their assembly. For instance, the DNA tetrahedron was readily constructed by the group of British and Dutch scientists just from the four component oligonucleotides each run around one tetrahedron face and

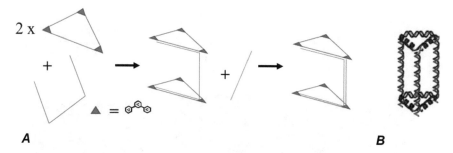

Fig. 3.6 Diagrams of the DNA nanoprism assembly using planar triangle modules. (**a**) Two single-stranded DNA triangles with vertices made of rigid organic molecules (shown as ▲) combined first with linking DNA strand (shown as ⊐) and then with rigidifying DNA strand (shown as /; to form rigid edges from duplex DNA) will self-assemble into DNA nanoprism shown schematically in (**b**), if three linking strands and three rigidifying strands will be employed. (Adapted from [15])

hybridized to form the double-helical edges [13]. Adjacent edges were connected at vertices through unpaired bases acting as "hinges" (Fig. 3.5). DNA tetrahedron formed in this way contains single-stranded breaks (or nicks), which are closed by DNA ligase.

Another facile method for assembly of diverse discrete 3D DNA nanostructures was developed by Canadian scientists who employed the toolbox of single-stranded cyclic 2D DNA building blocks representing triangles, squares, pentagons, and hexagons that contain rigid organic molecules as their vertices [15]. These plane polygonal blocks will serve as the faces or sides of the nano-objects to be constructed when they will be linked with auxiliary DNA single strands that form edges of these objects (see Fig. 3.6). Using this method, a number of polygonal prisms, heteroprisms, and biprisms having edges made of DNA duplexes were designed.

While making various assemblies from branched DNA junctions, researchers soon realized that flexibility of branched DNA junctions, which facilitates the construction of finite-sized, discrete 3D DNA objects, substantially limits, in contrast, their assembly into extended 2D networks—it can be seen from micrographs in Figs. 3.3 and 3.4, where 2D meshes of only small sizes were assembled from branched DNA junctions. To circumvent this problem, the alternative rigid plane DNA building blocks containing two crossover DNA junctions between helical domains at the two proximate branch points were designed in late 1990s by Nadrian Seeman and Erik Winfree with coworkers [16, 17]. These multistrand DNA double-crossover complexes can be viewed as the two juxtaposed planar four-arm junctions arranged in such a way that at each junction the non-crossover strands are antiparallel to each other, or as a flat tile comprising the two neighboring and aligned DNA double helices joined by single strands that begin on one DNA helix and switches onto an adjacent helix (Fig. 3.7a, b).[2]

---

[2]There are five possible DNA double-crossover motifs, but only two of them featuring antiparallel orientation of their two double-helical domains shown in Fig. 3.7 are stable.

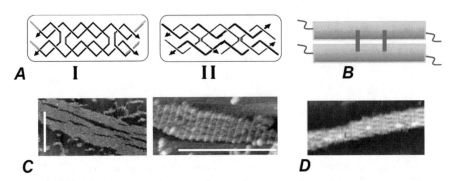

Fig. 3.7 Nano-assemblies of planar DNA tiles. (**a**) Schematic drawings of the two different structural arrangements of stable DNA double-crossover structures consisting of a pair of aligned double helices fused by strands that cross between them to tie them together. DNA double-crossover **I** is composed of only four DNA single strands, whereas DNA double-crossover **II** requires five DNA single strands for its construction. The arrowheads at the ends of the strands designate their 3′-ends. The crossover points are highlighted in red; possible sticky ends in DNA double-crossover **I** are in blue. The contour lines around double-crossover structures are drawn to highlight their tile-like shapes seen in micrographs. (Adapted from [16]). (**b**) 3D model of double-crossover DNA tiles featuring the molecular construct of two double helices (shown as cylinders) connected at two crossovers by "staple" DNA strands (blue lines) and terminated with sticky ends (zigzag lines). (**c**) Micrographs of a large periodic network self-assembled from the DNA tiles of type **I** with sticky ends, with each tile being ~2 nm thick and ~4 × 13 nm in surface size. 2D crystalline lattice of multiple tiles is clearly seen at higher magnification in the second micrograph, whereas the first micrograph shows that such a lattice is a monolayer of DNA built of thousands of DNA tiles. Both scale bars are 300 nm. (Adapted from [17].) (**d**) Micrograph of a part of the DNA nanotube with ~25 nm diameter self-assembled from the DNA triple-crossover tiles, which contain three coplanar double helices linked at each of the four crossover points. Thus formed nanotubes have been observed with lengths up to 20 μm. (Courtesy of Thom LaBean, North Carolina State University)

Based on reasonable assumption that elongated flexible molecules could readily form ring structures, whereas rigid molecules could not, negative results of cyclization studies with DNA double-crossover complexes proved that they in fact are very rigid constructs and, therefore, they should be the useful modules for assembly of long periodic networks [17, 18]. Indeed, by using DNA double-crossover structures as tiles, it was possible to assemble from them the 2D monolayered crystalline DNA lattices as large as several microns in size (Fig. 3.7c).

And it was done without the involvement of DNA ligase because these lattices are kept stable solely by noncovalent interactions, such as numerous hydrogen bonds and base stacking. This important feature substantially simplifies not only the assembly of these DNA lattices, but also their disassembly, as it could be necessary in certain applications (e.g., for drug delivery or in DNA computing; see Sect. 3.3 below).

To further extend the sets of building blocks for nano-assemblies involving DNA crossover modules, triple- and quadruple-crossover tiles with sticky ends have been constructed in 2004–2005 by the research groups from Duke University and University of Southern California [19, 20]. These larger DNA tiles consist of three or four aligned coplanar DNA double helices held together in a plane via two crossovers between each double helix (so that each triple- or quadruple-crossover tile is connected by a total of four or six crossovers, respectively). Such an upgrade resulted in more versatile uses: besides their self-assembly into diverse periodic 2D lattices, these new tiles were successfully wrapped into 3D nanostructures [21], such as DNA nanotube shown in Fig. 3.7d. Nonplanar multi-helix tiles (aka bundle DNA tiles) based on cyclic DNA crossover complexes have been designed around the same time, too, by the US academic research teams [19, 22]. These constructs provide additional avenues for tiling and folding into the third dimension (see Fig. 3.8).

Surprisingly, but in 2008 Peng Yin with other Caltech and Duke University scientists has demonstrated that certain 2D and 3D shapes can also be simply assembled from plurality of short DNA single strands, which all have no prior secondary structure while behaving like tiles during the assembly directed by them [23]. Their study extended the toolbox of building elements composed of single-stranded oligonucleotides: in the previously developed sets described above, each component oligonucleotide binds up to three other elements to form the discrete, finite-sized DNA objects (see Figs. 3.5 and 3.6), whereas in a new set each single-stranded DNA building block is capable of binding another four matching blocks (see Fig. 3.9). And this innovative feature makes a dramatic difference—the novel DNA blocks are capable of self-assembly into protracted nanostructures, such as DNA ribbons and nanotubes (Fig. 3.9). Unlike a multistranded crossover tile, which is a well-folded, compact structure displaying several sticky ends, a single-stranded DNA tile (SST) is a floppy construct composed entirely of concatenated sticky parts that folds into a rectangular shape because of its interaction with neighboring SSTs during assembly. Yet, the rigid DNA double helices formed by interactions between the

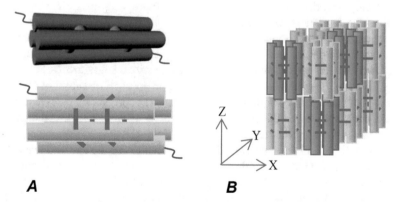

Fig. 3.8 Schematic representation of bundle tiles designed from DNA. (**a**) Four- and six-helix DNA bundles drawn as the two nonplanar composites of aligned cylinders symbolizing double helices formed from complementary DNA strands. The crossovers connecting DNA helices and shaping them into a bundle are shown by blue lines; zigzag lines represent sticky ends used in subsequent nano-assemblies. (**b**) A variety of such DNA bundles with sticky ends (not shown here) can self-assemble into various 3D structures, like shown here, i.e., hypothetical rectangular cuboid composed from the three different six-helix bundles (colored by yellow, blue, and green)

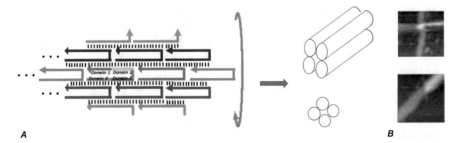

Fig. 3.9 Nano-assemblies with single-stranded DNA tiles. (**a**) Single-stranded DNA pieces employed in these assemblies comprise four recognition domains linked to one another, aka con-catenated, as shown up for one of them. This attribute makes it possible for each unit to pair with four other DNA units with complementary domains by making U-turn and configuring itself into a tile-like structure (arrows are for strand polarity; short vertical bars represent base-pairing). When mixed together, hundreds of single-stranded DNA tiles self-assemble in one-pot reactions into desired target structures mediated by inter-tile binding interactions. One kind of assemblies could be a variety of laterally extended DNA nanoribbons composed of multiple aligned DNA double helices. Another option is to transform a nanoribbon into a nanotube by merging its edges. This will convert a four-helix DNA ribbon shown here into a four-helix DNA tube; side view and cross-section view of a few aligned DNA nanotubes are shown schematically on the right. (**b**) Micrographs of segments of four-helix (top) and eight-helix (bottom) long nanotubes assembled from single-stranded DNA tiles and having diameters of about 13 and 23 nm, respectively. (Courtesy of Thom LaBean, North Carolina State University)

neighboring tiles result in emergent stiffness along the nanostructure's growth direction.

A few years later, Peng Yin, now at Harvard University Molecular Systems Lab, came up with a creative idea on how to employ single-stranded DNA tiles for making not only nanoribbons or nanotubes but also more complex DNA shapes, thus actually building any shape from DNA you can imagine. He realized that binding interactions between different tiles could be devised in such a way that plurality of them might potentially self-assemble into a square field called molecular canvas [24]. Then, any desired 2D shape, e.g., a rectangle with a hole shown schematically in Fig. 3.10, can be created simply by selecting an appropriate subset of DNA tiles from the common pool constituting molecular canvas and mixing them together. In a sense, each DNA tile in this pool acts as a pixel on a laptop's screen when a set of bright pixels forms digital image.

The simplicity and modularity of this approach allowed the Harvard researchers to build more than a hundred of distinct shapes from a master strand collection constituting 310-pixel molecular canvas. They include all letters of Latin alphabet, all digits of Arabic numeral system, numerous symbols, and emoticons, some of which are shown in Fig. 3.10. To automate painting with DNA, the team designed a software and a robot to pick the selected tiles. The desired shape is drawn using a graphical interface, and the robot picks out and mixes the required DNA strands. In such a way, it can produce a dozen of various shapes by working a 12-h shift without human intervention.

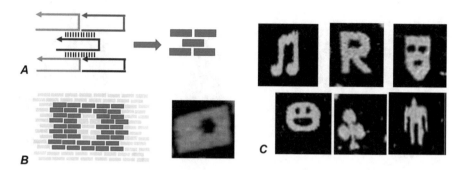

Fig. 3.10 The concept of molecular canvas and its realization. (**a, b**) A collection of diverse single-stranded DNA tiles can self-organize into a square lattice of parallel double helices connected by single-strand linkages, thus forming a miniature brick wall-like pattern that serves as a virtual molecular canvas. In fact, any desired shape, e.g., a 55 × 100 nm rectangle with a hole shown in (**b**), can be produced by one-pot annealing of an appropriate subset of DNA tiles (shown in blue), with the remaining tiles (shown in gray) being excluded. (**b, c**) Micrographs of a variety of shapes "drawn" by Peng Yin's team using molecular canvas formed by hundreds of short synthetic DNA strands folded into 3 × 7 nm tiles). (Adapted from [24])

Next, Peng Yin's team of scientists at Harvard has taught synthetic DNA a new trick: by modifying the design of single-stranded DNA modules they transformed flat DNA tiles into DNA bricks, thus extending modular assembly method from 2D to 3D. Like a single-stranded DNA tile, DNA brick has four binding domains, two head and two tail ones, designed to interlock with other DNA bricks by forming complementary double-helical connections. But in DNA tiles each binding domain consists of ten nucleotides and it forms full helical turn when binding a partner—that particular arrangement keeps the entire structure flat. A new trick is that in DNA bricks binding domains consist of only eight nucleotides so that they form nearly 3/4 of helical turn with complementary interlocking domain of other DNA bricks to be connected at right angles. Such a simple alteration lets the assembly of single-stranded DNA modules to break out of the plane into the third dimension [25].

The analogy with interlocking Lego bricks or Pinblocks helps to understand how DNA bricks work if we will model them with plastic building blocks featuring two protruding round plugs to represent two tail domains and two recessed round holes to represent two head domains. The two such pieces can interlock at a right angle by insertion of a plug into a hole, which will symbolize head-to-tail hybridization between a pair of DNA bricks carrying complementary sequences (see Fig. 3.11a, b). Then, a variety of 3D shapes can be built by connecting a number of these plastic building blocks in a requisite order and so this was done with DNA bricks: the Harvard team of scientists constructed numerous distinct assemblies exhibiting sophisticated surface features, as well as intricate interior cavities and tunnels composed of hundreds of DNA bricks [25].

Recently, the Peng Yin's team has tested the second-generation DNA bricks, which have longer binding domains to make the entire structure more stable. As a result, larger DNA constructs with complex architectures have been assembled,

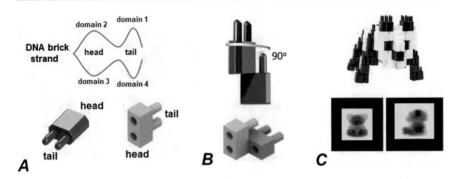

Fig. 3.11 Sculpting with DNA. (**a**) Strand model of a DNA brick (top) and its plastic block mimics—Pinblock and Lego brick (bottom). (**b**) A pair of the two plastic blocks interlocked at a right angle. (**c**) Near 10-cm-tall 3D figurine of a panda bear constructed from black-and-white Pinblocks (top) and micrographs of the front and side views of million-times-smaller teddy bear shapes assembled from DNA bricks by using similar to Pinblock's interconnection principle. (Bottom; adapted from [26])

including a bunny and a teddy bear (see Fig. 3.11c), which were sculpted of thousands of DNA oligonucleotides [26]. The largest structure built of upgraded DNA bricks is DNA cuboid having more than 100 nm in sizes and about 1 GDa in weight (equivalent to a pack of 1000 million hydrogen atoms), and containing tens of thousands of building blocks. To realize how big is this artificial DNA construct, be aware that a single particle of Rous sarcoma virus weighs 0.3 GDa and that of herpesvirus weighs 2.7 GDa!

A large assortment of DNA construction sets described above, as well as the amazing variety of tiny objects that can self-build from them—all this prompts me to consider them as the most amazing building sets ever known. But besides playing with DNA construction sets, scientists also invented a different game for sculpting with DNA called DNA origami, which is a new twist in the story of DNA self-assembly described below.

3.2   DNA Origami: Another Robust Tool for DNA Sculpting and Painting

Origami (from Japanese *ori* meaning "folding," and *kami* or *gami* meaning "paper") is the art of paper folding originated in Japan. The aim is to create some figurine from a single flat piece of paper by making multiple folds. And kids love to construct simple origami shapes, like paper boat and paper airplane (Fig. 3.12a), to let them float in the water or fly in the air. Now scientists are playing similar game at nanoscale with long single-stranded DNA molecules by folding them into the desired 2D and 3D shapes.

DNA origami originates from the pioneering studies published by the Scripps and Caltech scientists in the mid-2000s. First, in 2004 William Shih with two colleagues from the Scripps Research Institute managed to fold a 1.7-kb (kilobase

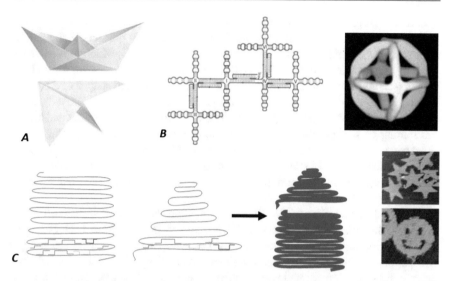

Fig. 3.12 Origami: the art of folding. (**a**) Origami boat and origami airplane, each folded out of a piece of paper. (**b**) Folding of DNA octahedron. Left: Secondary structure of the branched-tree folding intermediate. The structure consists of a single long backbone strand (black) and five auxiliary strands (blue) that form stems in this structure by binding to backbone strand. Right: Combined 3D image of the 20-nm-sized DNA octahedron constructed from multiple micrographs of individual DNA octahedral nanostructures seeing at different 2D projections; see the right-bottom insert showing the two of them. (Adapted from [27]) (**c**) Left: Schematics of creating rectangular and triangular DNA origami shapes by raster-filling the desired shape with long single-stranded scaffold DNA and by securing it with multiple short oligonucleotide staple strands (only few of them are shown in different colors for simplicity). Right: Micrograph of several star pentagon shapes and micrograph of a disk with three holes resembling a smiley face emoticon, all created by DNA origami technique. (Courtesy of Paul W.K. Rothemund, California Institute of Technology)

pairs)-long artificial DNA single strand into a nanoscale octahedron (Fig. 3.12b). The newfangled 3D DNA construct was built with the aid of five short auxiliary oligonucleotide strands, which helped to fold the longer DNA strand containing a specially designed sequence of nucleotides into the branched-tree folding intermediate (Fig. 3.12b). Then, this intermediate structure spontaneously self-associated via its conjugate terminal branches to form the ultimate octahedral shape [27].

This proof-of-principle study has demonstrated that the folding of long DNA strands could proceed in a correct predesigned manner by efficiently avoiding potential misfolds. So, it suggested a new approach for making DNA nanostructures, but, unfortunately, it lacked the generality. Indeed, the sculpting of lengthy DNA strands into diverse shapes similar to how it was done with DNA octahedron would, in principle, require quite a tricky design of a variety of long artificial DNA strands with different nucleotide sequences, each of which suits only particular DNA nanostructure.

Yet, two years later, encouraged by the potent idea of sculpting long DNA strands into designer nanoscale architectures, Caltech's Paul Rothemund came up with the general robust principle on how to fold the same long DNA strand into the limitless variety of different shapes. Rothemund smartly envisioned that any shape can be approximated by folds of a single long "scaffold" DNA strand, which runs back and forth in a raster-like pattern similar to a movement of electron beam on a television screen or computer monitor (see Fig. 3.12c for the raster-filled shapes of a triangle and a rectangle). He also realized that special sets of oligonucleotide "staple" strands, which bind to specific positions on the scaffold, can both direct and secure the folding of scaffold DNA strand into predetermined shapes by forming cross-links between raster lines of DNA scaffold (aka DNA frame). Moreover, binding of staple strands to scaffold strand can be arranged in such a way that each DNA raster line will form essentially continuous double helix, thus making the final structure more rigid and dense.

Using all these rather simple tricks, Rothemund was successful in sculpting a 7-kb-long linearized single-stranded DNA scaffold isolated from bacterial virus (aka bacteriophage) M13 into several distinct symmetric 2D shapes, including a five-pointed star and a smiley face emoticon shown in Fig. 3.12c [28]. Remarkably, every other shape was formed with a different set of over 200 short oligonucleotide staple strands but with the same M13 DNA scaffold strand, similar to creating different paper figurines by differently folding identical pieces of paper with original origami techniques (see Fig. 3.12a).

This inventive study gave rise to the scaffolded DNA origami methodology, another versatile tool for constructing with DNA, which was rapidly extended from building various 2D DNA structures of both symmetric and asymmetric desired shapes[3] into the third dimension by employing various adaptations of the original Rothemund's technique. In particular, the group of Danish and German researchers completed in 2009 the study, which resulted in DNA origami design of a nanoscale DNA box with a closed lid that can be opened by externally supplied DNA "keys" [30]. This box was self-assembled from circular M13 DNA scaffold strand by annealing it with 220 staple strands to form 6 interconnected rectangular sheets as faces of the box. These sheets were subsequently 3D arranged into a cuboid box of $42 \times 36 \times 36$ nm^3 in size by using 59 additional staple strands that fasten the sheets' edges together (Fig. 3.13a, b).

Remarkably, in this nanoscale 3D DNA origami structure one of the six rectangular sheets was fastened to a box only via a single edge, thus making a lid with "hinges" composed of scaffold linkers (Fig. 3.13b). The box has initially been held closed by a "double lock" formed by two short DNA strands protruding from the lid and annealing with two complementary strands protruding from the main box. Each strand of the lock has also a small sticky end where the "key" oligonucleotides, when added to a box, can bind and open the lock by strand displacement. Micrographs of DNA boxes with unlocked lids proved that they subsequently became to be open (see Fig. 3.13c), and additional evidence of the lid opening was independently

---

[3] See for instance an article of Danish scientists reporting the self-assembly of asymmetric 2D DNA origami structures having the shape of a dolphin [29], which makes up a central part of the seal of Aarhus University (Denmark).

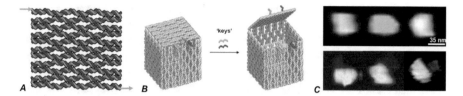

Fig. 3.13 Design of a DNA origami box. (**a**) Molecular model of one of the six DNA sheets obtained by folding of DNA scaffold strand (shown in gray) with multiple staple strands (shown in deep blue); connections of this sheet with two neighboring similar sheets are marked by arrows. (**b**) A model of the DNA box with a lid assembled from the sheets shown in (**a**) by using additional staple strands that fasten the sheets' edges together. This box is held closed by the two "locks" (orange and blue) that are double helices formed by two short strands protruding from the lid and the main box, respectively. The box can have its lid open if locks are unlocked with the two matching "key" oligonucleotides by strand displacement. The reporter system attached to DNA boxes changes its color from red to green upon opening a box's lid. (**c**) Micrographs of the boxlike DNA origami nanostructures with lids closed (top) and open (bottom). (Adapted from [30])

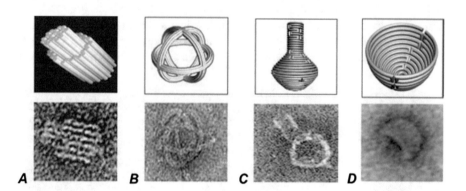

Fig. 3.14 Design of curved DNA origami shapes. Top row: software-generated models of nanostructures with cylinders and rods representing DNA double helices to be formed and packed by DNA origami via attaching multiple short staple DNA strands to a single long DNA scaffold strand; bottom row: micrographs of thus created sub-100-nm-sized DNA origami nano-shapes that look like a bottle with neck (**a**; adapted from [31]), a beach ball (**b**; courtesy of William Shih, Harvard Medical School), a flask (**c**), and a bowl (**d**; both courtesy of Dongran Han, Arizona State University). Structures **a–c** were obtained with the full-length M13 DNA scaffolds, whereas structure D was folded from a half-length M13 DNA scaffold

obtained from a specially designed reporter system attached to DNA boxes, which changed the color of emitted light upon opening the box lids.

Besides simple geometric shapes, like the rectangular DNA box described above, two groups of researchers, one from Harvard University and another from Arizona State University, managed to design in 2009–2011 a variety of 3D DNA origami curved shapes, both concave and convex ones [31–33], some of which are shown in Fig. 3.14. All these nonplanar 3D nanostructures were designed with the aid of specialized graphic software tools enabling the researchers to build the intended DNA origami structures on the computer screen helix by helix, thus designing the folding path, crossover pattern, staple positions, and a list of staple sequences.

Another advance in the DNA origami techniques has been recently made by the group of Swedish and Finnish researchers who developed in 2015 a new elegant computerized strategy allowing to fold DNA strands into much more complex structures than those obtained before [34]. Unlike previous software programs, the starting point of a new method is a 3D polygonal digital mesh designed in the computer and approximating the spatial geometry of an object one wishes to realize at the nanoscale (see Fig. 3.15a). Then the newly developed software directly provides researchers with all the necessary details without any manual intervention, i.e., which staple DNA sequences should be generated and how to organize the scaffold and staple DNA strands along the edges of a hollow mesh to make the desired shape in DNA that looks like this mesh on a computer screen.

Thus, the DNA origami design can be performed in such a completely automated fashion, with a recipe for preparing a particular DNA structure being given directly from digital 3D mesh, including a list of the ingredients required. Several complex hollow shapes, including 50-nm-sized bunny (Fig. 3.15), have been generated this way. Note that dissimilar from conventional origami designs usually built from close-packed bundles or stacks of helices, as it can be seen from Figs. 3.13 and 3.14 (with the exception of a DNA nanostructure shaped as a beach ball), the new method

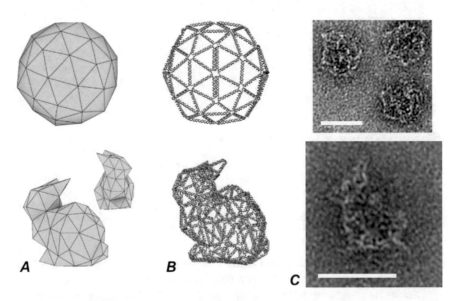

Fig. 3.15 DNA origami design based on polygonal meshes. (**a**) Software-generated models of hollow nanostructures drawn as the 3D triangulated meshes and representing icosahedral ball (top) and a bunny figurine (bottom). (**b**) DNA origami models of the respective polygonal meshes with their edges replaced by DNA double helices that are formed via binding of a single long DNA scaffold strand (green) to multiple short staple DNA strands (blue). (**c**) Micrographs of the corresponding DNA origami structures assembled with the derivative of M13 DNA scaffold. Scale bars are 50 nm. (Adapted from [34])

endowed nanodesigners with more open nanostructures, which instead use isolated double helices as structural elements.

A variety of DNA origami techniques presented above enable the bottom-up fabrication of diverse DNA nanostructures by designing hundreds of staple strands with complementary sequences to the specific binding locations of a scaffold strand. Together with the miscellaneous DNA nanostructures that can be built by using DNA construction sets, this extends further the robust DNA potential as a nanoscale building material. And I think now it is time to explain below to my readers why numerous researchers are so compelled and captivated with building the DNA-based nanoconstructs, besides all the aesthetic pleasure they evidently get by doing this.

3.3 Values of the Folds: DNA Nanostructures in Material Science and Technology

Even turpentine oil is useful for something!

While throwing stones into the water, look at the circular waves they form. Otherwise, this will be just an idle entertainment.

Kozma Prutkov (Russian writer), Fruits of Reflection: Thoughts and Aphorisms
(1853–1854)

Yes, indeed, after reading the previous sections, you, my reader, may wonder what are all these nanoscopic DNA-based constructs and nanoscale meshes made for? Why scientists spend their time and efforts for creating nano-sized DNA boxes, tiles, and gridirons, or nano-flasks and nano-bowls made of DNA, not to mention too-small-to-see smiley faces, bunnies, and teddy bears? And the answer is this: the studies described above were not just an idle entertainment performed for impractical self-amusement (rephrasing the wise Prutkov's words written at the beginning of this section). They all have been done for the proof-of-concept demonstrations of what came to be called structural DNA nanotechnology [3], the ultimate goal of which was, and still is, to discover the best novel ways of making things in different areas—from medicine and biosensing to computing and robotics.

3.3.1 DNA Nanostructures in Medicine and Biosensing

The first thing that may come to mind is to employ hollow DNA nano-constructs as tiny capsules for delivery of encapsulated drugs. Indeed, in contrast to unstructured DNA, which is readily degradable by various nucleolytic enzymes (aka nucleases) of bodily fluids, DNA nanostructures are quite resistant to enzymatic digestion due to their overall closed and compact shapes [35]. Therefore, hollow DNA nano-constructs can be used for targeted delivery of encapsulated drugs, sometimes called smart drug delivery [36], to increase the concentration of the injected medication in chosen parts of the body relative to others. For that purpose, DNA nanocapsules should be programmed to target specific locations or cell types in the body and then to release the encapsulated drug, thus minimizing possible side effects of drugs to the surrounding tissues [37].

Although clinical applications of these innovative drug vehicles are still far from realization, several pilot studies employing them for delivery of various medications to specific cells of different animals, such as cockroaches, roundworms, and mice, have already been conducted with promising results.

For instance, DNA tetrahedrons constructed in 2005 from a set of oligonucleotides by British and Dutch scientists [13], as it was described above, have been successfully used later by others as molecular cages for targeted delivery of anticancer drugs to malignant tumors in mice. Indeed, in 2012 the combined group of American researchers from academia and pharma industry reversibly loaded onto DNA tetrahedrons small interfering RNAs (siRNAs)—a therapeutic agent that suppresses the expression of targeted genes—to retain the drug in blood for prolonged duration for maximum absorption and better bioavailability, as compared to the naked siRNA [38].

To selectively target cancer cells, siRNAs were conjugated with the folate molecules and the corresponding DNA nano-constructs were then injected to mice carrying malignant tumors overexpressing folate receptors (Fig. 3.16). It was found in this study that the blood circulation time of DNA tetrahedron-loaded siRNAs has increased fourfold, that these nano-constructs were readily taken up with their cargo by cancer cells, and that they caused the silencing of target gene expressed in tumors, which could potentially block their proliferation.

Four years later, Chinese research team from Jinan University reported the reduced tumor growth in mice administered with the repeated doses of DNA tetrahedrons loaded by small-molecule anticancer drug RuPOP [39]. The drug was found to be accumulated in tumors and it did not exhibit any general toxicity even at the relatively high dose used in this study.

Researchers from Purdue University employed self-assembled DNA polyhedrons as frames for assembly of highly symmetric DNA-protein complexes resembling natural viral particles in both size and shape (Fig. 3.17a). They also suggested

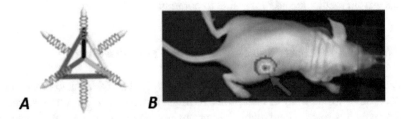

Fig. 3.16 Targeting malignant tumor in mice with siRNA-loaded tetrahedral DNA nanocapsules. (**a**) Schematics of DNA tetrahedron noncovalently attached to six siRNA molecules by their hybridization to tetrahedron's edges (shown in different colors). Folate ligands are drawn as bullets on the ends of siRNA strands. (**b**) Exemplary image of the nude mouse obtained by fluorescent photography and demonstrating targeted accumulation of DNA nanocapsules at the tumor site (pointed by red arrow) after injection of fluorescently labeled nano-constructs to tumor-bearing animal. (Adapted from [38])

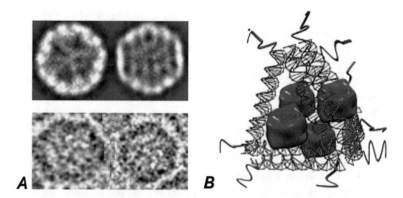

Fig. 3.17 Artificial viruslike nanoparticles assembled on polyhedral DNA frames. (**a**) Micrographs of 20-nm-diameter nanoparticles (top image) comprising DNA icosahedron covered by nearly two dozen of the DNA-bound streptavidin proteins. They look similar to icosahedral viral particles of bufavirus (bottom image)—human pathogen causing gastroenteritis and acute diarrhea, which is also about 20 nm in diameter and has five dozen capsid protein molecules of the size of streptavidin. (Adapted from [40, 41].) (**b**) Schematics of viruslike nanoparticle vaccine assembled on the tetrahedral DNA frame (green helices). Four streptavidin proteins bound to DNA and shown in red serve as model antigens; the DNA tetrahedron-bound CpG oligonucleotides depicted as purple ribbons act as representative adjuvants. (Adapted from [42])

that such artificial nano-constructs decorated with different immunostimulatory molecules would serve as potent synthetic vaccine, which induces much stronger immune responses than individual immunostimulants [40].

Motivated by this bright idea, the Arizona State University scientists used DNA tetrahedron as a platform for assembly of vaccine nanoparticles formed by the antigen-adjuvant-DNA complexes (Fig. 3.17b). When they immunized mice with injections of this man-made "virus," they found that compared to an unassembled mixture of antigen and adjuvant, the assembled composite nanoparticles induce stronger and longer lasting antibody responses against the antigen without any immune reactions to the DNA nanostructure itself [42].

Recently, icosahedral DNA nanocapsules loaded with the neurosteroid dehydro-epiandrosterone, which promotes neurogenesis and neuron survival, were used by the scientists from the University of Chicago to deliver this drug to cultured mouse brain neurons in a controlled way via light-triggered release [43]. For doing this, the drug was conjugated to synthetic dextran polymer via photocleavable linker. Then, the conjugates were inserted in the DNA icosahedron by mixing them with icosahedron halves self-assembled from five-arm DNA junctions, and the halves were connected to each other upon mixing, thus encapsulating the drug and forming the complete DNA nanostructure. After adding these nanostructures to nerve cells, they were irradiated with the violet light to release a drug from its polymeric carrier, which allowed the diffusion of free drug out of the nanocapsules and subsequent activation of those neurons on which the nanocapsules are localized.

A few years earlier, Indian researchers employed DNA icosahedrons as nano-delivery vehicles of fluorescently labeled dextran for functional in vivo imaging in roundworms (aka nematodes)—one of the first demonstrations of delivering the drug cargo by DNA nanocapsules in live animals [44]. They showed that when injected in worms, the DNA-caged fluorescent drug was directed to worm's specific immune cells (called coelomocytes, a type of leukocytes functionally analogous to some human white blood cells), which are known to express the DNA-binding receptors on their surface. In contrast, the injected uncaged drug was widely distributed within the animal body (Fig. 3.18).

Besides polyhedral DNA nanocapsules, DNA origami structures with sizes of a few 100 nm have also been tested in mice for caged drug delivery. Compared to the DNA polyhedrons that are about 10–20 nm in size, they are much larger with respect to their mass and charge, which imposes certain obstacles for their cellular uptake. But they may accumulate in the tumor region due to the well-known feature of malignant tissues to become permeable for large molecules and small particles that is called enhanced permeability and retention (EPR) effect [45].

Accordingly, Ding and coworkers from Chinese Academy of Sciences and National Center for Nanoscience and Technology have found that administered to mice triangular shaped DNA origami nanostructures with the 120-nm-long sides indeed accumulated in the tumor region [46]. Then, they loaded these nanostructures by anticancer antibiotic doxorubicin, which intercalates DNA duplexes at physiological neutral pH, and injected them to tumor-bearing mice. Since most tumor tissues have acidic environment in contrast to normal tissues, nanostructures released the payloads at tumor site because doxorubicin-DNA complexes become less stable under these conditions. As a result, the antitumor efficiency of drug-loaded DNA origami nanostructures was significantly higher compared with free doxorubicin.

In a next step, these researchers used the DNA origami triangles for fighting cancer with light via delivering with their help the rodlike gold nanoparticles directly to mouse tumors [47]. By applying near-infrared laser irradiation, the tumor-targeted gold nanorods can be specifically heated and thus selectively kill tumor cells.

For triggered release, the barrel-shaped closed nanocapsules have been designed via collaboration of Israeli and American scientists using DNA origami nano-constructs with a gate that opens in response to the presence of specific proteins [48]. This opening will expose the molecular payload hidden inside, thus making it

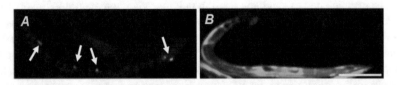

Fig. 3.18 DNA-encapsulated fluorescent drug is targeted to specific cells of a tiny soil-dwelling nematode *C. elegans*. Representative images showing (**a**) a worm injected with a solution of DNA icosahedrons loaded with fluorescent drug and (**b**) another worm injected with the uncaged drug. Arrows in (**a**) point to a few coelomic cavities occupied with coelomocytes, which selectively uptake the nanoparticles delivering fluorescent drug; scale bar in (**b**) is 10 μm. (Adapted from [44])

available to engage target cells. Indeed, when these DNA nanocapsules were loaded with antibodies recognizing the insect's blood cells and injected in living cockroaches, the antibodies attacked blood cells only when nanocapsules were co-injected together with "key" proteins.

Thus, in real-life therapeutic treatments, this strategy could be exploited, e.g., to selectively target malignant tumors when malignancy markers will be used as the "key" to open the drug-loaded nanocapsules attached to tumor cells. Most intriguingly of all, the Israeli research group has recently demonstrated that these nanostructures behaved like nanoscale robots in a living host, which could be remotely controlled by human brain activity (see the next chapter of this book devoted to a variety of DNA-based machines and robots, including DNA origami nanorobots).

As to biosensing applications of DNA nanostructures, an elegant example was developed by Dan Luo and coworkers from Cornell University who assembled the dendrimer-like spherical DNA nanostructures from the set of fluorescently labeled three-way DNA junctions to generate a multiplexed detection system [49, 50]. This

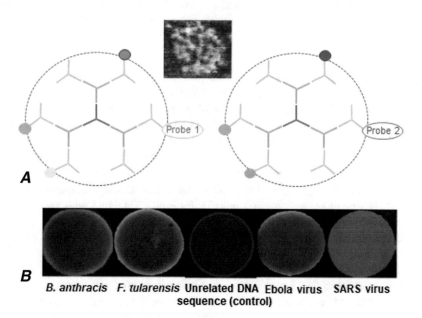

Fig. 3.19 Multiplex biosensing with dendrimer-like spherical DNA nanostructures. (**a**) Schematics of DNA dendrimers self-assembled from the Y-shaped three-arm junctions with sticky ends shown in the three shades of blue to distinguish between successive layers of dendrimer structure (dotted circles are used to highlight the dendrimers' circular shapes clearly seen in the micrograph inserted above). Dendrimer on the left will recognize the corresponding analyte by probe 1, whereas dendrimer on the right will recognize a different analyte by probe 2. Dendrimers with different probes are labeled by tags with different colors (or combinations of differently colored tags, as it is shown here) to selectively visualize distinct probe-analyte complexes. (**b**) DNAs from four deadly pathogens along with control DNA spotted onto nylon membrane and detected simultaneously by a set of four DNA dendritic probes, which were labeled by four different colors and visualized by corresponding optical filters. (Adapted from [49, 50])

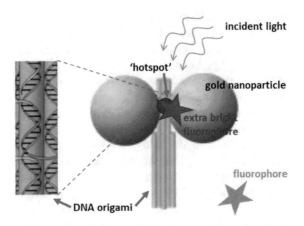

Fig. 3.20 Design of the DNA-based nanoantenna with nanolens. The nanoantenna is composed of a DNA origami column consisting of a 12-helix bundle and having a length of 220 and 15 nm diameter. Two 100 nm gold particles attached to it act as a nanolens which focuses incident light to a few nm "hot spot" located inside the 23 nm gap between nanoparticles. A fluorescent dye molecule positioned within the hot spot will be hundred times brighter than a dye molecule outside the hot spot

system consists of an assortment of near-30-nm-sized DNA spheroids, with each kind of them carrying a probe for specific pathogen and a distinct combination of fluorescent dyes serving as an optical code for particular probe (Fig. 3.19). It was shown by Cornell researchers that the DNA diagnostics developed by them was indeed capable of rapid simultaneous detection of multiple pathogens, and that such an analysis could identify as low as a thousand bacterial or viral particles or even less in just a milliliter of biofluid [50].[4]

Another biosensing application of DNA nanostructures was elaborated by the team of German nanobioscientists directed by Philip Tinnefeld who designed the DNA-based nanoantenna to increase 100-fold the fluorescence intensity of a dye molecule bound to it. This nanoantenna is composed of a pillar-shaped DNA origami column, to which a pair of gold nanoparticles is attached [51]. The closely located nanoparticles act as nanolens by focusing the incident light on the tiny hot spot between them, thus greatly intensifying illumination of a fluorescent molecule located within and consequently enhancing emission of secondary light (Fig. 3.20).

The team intends to exploit this technology for diagnostic purposes by boosting the sensitivity of optical biosensors comprising multiple such nanoantennas. To this means, a binding site for DNA or RNA markers of different diseases can be created in the hot spot. If a particular disease-specific marker, e.g. myotonic dystrophy-specific DNA marker [52] or Crohn's disease-specific RNA marker [53], is present in a test sample, it will bind to the hot spot. Then, a fluorescently labeled hybridization probe to DNA or RNA marker will be located within the hot spot, therefore

---

[4]This is really low amounts of microorganisms, as there are up to 100 million human symbiotic microbes in only 1 ml of our saliva.

generating a highly enhanced fluorescent signal indicative of the disease. In 2015, this promising project won second prize in nationwide competition of German startup companies.

3.3.2   DNA Nanostructures in Binary and Logical Computing

Outside of medicine and biosensors for medical applications, there are also prospects of using DNA in biomolecular computing by self-assembly. Leonard Adleman was the first who demonstrated that DNA self-assemblies could be used to solve mathematical combinatorial problems. In the 1994 study, he proved that 1D DNA assemblies, i.e., DNA duplexes formed by cross-annealing of oligonucleotides from specially designed sets of short DNA sequences, would mimic the sets of dots and lines connecting them, called graphs [54]. With these DNA constructs, Adelman was able to algorithmically compute the Hamiltonian path for any directed graph, that is, to find a route connecting a chosen pair of dots in a particular graph with the "one-way" connections which go through every other dots only once—a mathematical game invented in the nineteenth century by Irish mathematician William Rowan Hamilton.

This groundbreaking work establishing the first experimental connection between DNA self-assembly and computation by programming DNA molecules to manipulate themselves inspired in the mid-1990s Erik Winfree, then PhD student at Caltech, to extend this methodology on the 2D DNA crossover tiles, which he assembled at that time in collaboration with Nadrian Seeman (as described in Sect. 3.1; also see ref. [16]). Winfree envisioned that these tiles may serve as symbolic representations of binary digits and binary logical operators/functions,[5] and they could be programmed for algorithmic self-assembly of complex patterns so that to serve, in principle, as the basis for simple logical operations [55, 56].

Winfree's idea was built on the principles of computing with interactive tiles, which were introduced in the early 1960s by Chinese mathematician Hao Wang, who originally called them as a finite set of square plates [57, 58]. The set of Wang tiles comprises a number of four-sided flat pieces with different shapes or different colors on each side so that these pieces can be joined together at complementing sides like a jigsaw puzzle, thus forming distinctive arrays that simulate basic logic computations [59]. An example of binary counting with seven kinds of jigsaw-shaped tiles is shown in Fig. 3.21a. In this example, each kind of tiles has protruded or indented edges of particular shape that match edges of other tiles, thus programming them to assemble into the table of consecutive binary numbers. In a similar way, a different set of computational tiles can be programmed by some simple hierarchical rules encoded in their edges to build a template, which performs a

---

[5] Because of its straightforward implementation in digital electronic circuitry using logic gates, the binary digit system of 0s and 1s is used by almost all modern computers and digital communications.

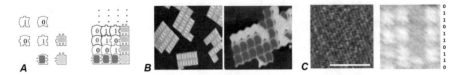

Fig. 3.21 Binary counting with tiles. (**a**) Theoretical example of counting with abstract Wang tiles. The basic set of component tiles is shown schematically on the left: it comprises three border tiles (shown in blue, green, and red colors) and four digital tiles marked with the values of 0 or 1. On the right it is shown aperiodic array assembled from these tiles. The array is composed of the reverse L-shaped border, which determines the 1 tile in its bend, then that 1 tile and the horizontal-border tile on its left determine the 0 tile that fits, while the 1 tile and the vertical-border tile above it determine another 0 tile that fits, and so on and so forth (as shown by dots). As a result, this array displays the table with binary numbers 001, 010, and 011, which correspond to their decimal counterparts 1, 2, and 3. (Adapted from [56]) (**b**) Experimentally obtained self-assembled mesoscopic periodic and aperiodic arrays of small plastic tiles floated in liquid and joined together by forces of surface tension and capillary action. Matching rules for these tiles were created by the laser-cut notches at their edges enforced by coating with patterns of hydrophobic and hydrophilic patches. (Courtesy of Paul W.K. Rothemund, California Institute of Technology). (**c**) Counting with the self-assembled molecular Wang tiles. Two types of equal-sized DNA tiles, i.e., "0" and "1" tiles, were prepared and self-assembled upon mixing together into the striped binary pattern seen in the micrograph on the left. "0" tiles are totally flat and are visible as gray rectangles, whereas "1" tiles have a bump in the form of attached DNA hairpins to provide a contrast and therefore are visible as white rectangles. Enlarged part of the assembly is shown on the right where the track of periodic "011" is clearly seen. Scale bar is 300 nm. (Adapted from [56])

computation by successive connection of specific tiles or encodes certain string of mathematical symbols into logic formulas [60].

Evidently, miniaturization and self-assembly of Wang tiles would make the computing by mathematical tiling to be readily amenable for automation. In an elegant study, Paul Rothemund, the pioneer of DNA origami, designed several sets of a few millimeter-sized floating plastic computational tiles, which were capable to encode simple binary functions by their self-assembly into patterned layers (see Fig. 3.21b) [61]. Yet, in Winfree's vision such an autonomous computing system can be made much more miniature in size by employing molecular tiles made of DNA. These microscopic tiles have four sticky ends (see Fig. 3.7) that might serve as the easy programmable binding domains connecting assorted DNA tiles. Indeed, the matching rules for DNA tiles can be readily encoded in the nucleotide sequences of sticky ends, which will algorithmically drive the assembly of nanosized tiles similar to that of four-sided macroscopic and mesoscopic Wang tiles.

The first attempt for real-life realization of this dream is shown in Fig. 3.21c, which demonstrates the proof-of-principle results published in Winfree's 2000 paper [56]. In there, two types of DNA tiles that differ in surface topography to represent the binary digits (either 0 or 1) when visualized by microscopy can be seen as arranged themselves in a self-assembled pattern of 0s or 1s.

The success of this pilot study, along with the pioneering work by Adleman, prompted a flurry of research activities, both experimental and theoretical ones, investigating the practical feasibility of computing with DNA. All these studies proved the robust ability of DNA self-assembly to perform sophisticated computations. Still, the estimated speed of DNA-based computation could be at best 10^{12} bit operations per second [62], which is hundred or more times lower compared to current silicon-based electronic computers performing 10^{14} bit operations per second [63] or emerging optoelectronic computers that could perform more than 10^{17} bit operations per second [64].[6]

That is why it doesn't look like DNA computers will be replacing conventional silicon computers any time soon. A more realistic application of DNA computing might be to employ DNA as software molecules in the DNA-based logic gates, i.e., molecular sized devices that perform certain logical operations on one or more inputs and produce single binary logic outputs, such as "YES"/"NO," "TRUE"/"FALSE," "ON"/"OFF," or "1"/"0"—the task, which is now primarily implemented using diodes or transistors acting as electronic switches. But notwithstanding the higher speed of computation achievable with silicon devices compared to DNA-based computers, the DNA however can do what electronic transistors cannot.

Indeed, in contrast to electronic logic gates inputted by electrical signals, the DNA-based logic gates are switches allowing both input and output information to be in molecular form, which let them serve as autonomous programmable "logical" controllers of designated biological processes [66]. To this end, a number of logic gates have been created using assemblies of DNA tiles and other DNA nanostructures, e.g., the DNA tile-based 1D and 2D assemblies of variants of exclusive OR (aka XOR) gates [67, 68] or the DNA origami-based 2D assemblies of AND gate [69]. As an example, Fig. 3.22 shows schematically the DNA-based design of 2-input logic AND gate and its transistor-based counterpart. Notably, in the latter work the group of Chinese researchers used the DNA-based logic gate AND for cardiac diagnostics based on the detection of two distinctive micro-RNAs [69], which are indicators of heart failure [70] and serve as well as inputting molecules for this 2-input logic gate. Such a molecular chip can perform simple computing and display correct symbols in response to disease indicators (see Fig. 3.22).

Furthermore, the DNA-based chips of that kind have the potential to not only logically analyze the biological microenvironment, but also activate or dispense inside the body a combination of drugs targeting a disease upon positive diagnosis deduced by them, as it was neatly demonstrated by Israeli researchers. They designed the multi-input logic gates of the stem-and-loop hairpin-like DNA nanostructures, which sense the level of expression of prostate or breast cancer oncogenes and other markers of these malignancies, and compute output either "NO" (for healthy state) or "YES" (for disease) [71, 72]. In case the output of such a diagnosis is "YES" (or "TRUE"), the DNA nanostructure transforms itself into a drug or releases the drug(s), which it carries.

---

[6]There are some thoughts, however, that nanosized DNA-based unconventional microprocessors would trade space for time. Then, a desktop DNA computer could potentially utilize more processors than all the electronic computers in the world combined, and thereby outperform the world's current fastest supercomputer, while consuming a tiny fraction of its energy [65].

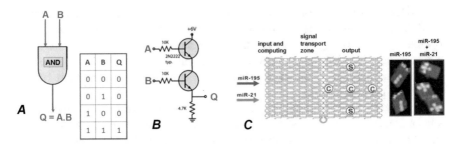

Fig. 3.22 Design of logic gate AND. (**a**) Schematics of the AND gate with the two inputs A and B, and output Q. The logical operation table (aka, truth table) for gate AND on the right shows all possible binary combinations of inputs to the gate with the resultant output for each combination of these inputs. This logic gate is named AND because, the output is "positive" when both inputs are "positive"; otherwise, the output is "negative." (**b**) Electronic diagram of the AND gate designed with two transistors. The electric current will flow through output terminal Q only if input terminals A and B are electrically active at the same time. (**c**) The DNA-based AND gate built from rectangular sheet of DNA origami as a support plate, onto which the input/computing and output modules were constructed from DNA, as well. The input/computing module contains immobilized strands of signal DNA, which could be released only if both analyte molecules, micro-RNAs miR-21 and miR-195, are present together in a test solution. The released signal DNA diffuses to output module, where it is captured by the two signal spots (marked S on the schematics), which generate signals when strands of signal DNA reach them and are labeled by the signal-producing molecule. The output module also has three control spots (marked C on the schematics), which generate signals by the signal-producing molecule itself. Micrographs on the right prove that thus designed DNA-based logic gate is "positive" in the presence of two micro-RNAs (i.e., a plus sign is displayed), and the output is "negative" when one of them is missing (a minus sign is displayed). The signal-producing molecule here is the streptavidin protein, which binds biotinylated signal DNA and is seen on the micrographs as white dots. The size of this molecular chip is about 100 nm. (Adapted from [69])

The researchers consider their developments as the first step towards a future of "smart" medical diagnosis and therapy in the form of "programmable" drugs. All this sounds like a science fiction, but maybe someday you will actually need no doctor to identify the disease and to prescribe you the treatment. Instead, when you will feel sick, you will go to local pharmacy to buy over-the-counter "magic" pills with DNA nanostructures acting as a computational cassette, which will both "diagnose" and "treat" you. Indeed, if DNA, in the form of genes, has the power to program the development of the entire healthy organism, why should we doubt the potential of some special DNA constructs to recognize the pathology and to fix it. And this future is not so distant: some experts think that within 5–10 years DNA-based digital logic circuits could be tested for medical applications.

Moreover, in a move to marry the bio-logic gates and microelectronics, another group of Chinese researchers constructed the DNA-based logic gates XOR on microelectrodes that functioned as a dual-analyte binary-like electrochemical sensor for ATP and DNA detection [73]. This smart development, along with the ability of DNA-based digital logic circuits for signal restoration, amplification, feedback, and cascading [74], would significantly facilitate the use of DNA logic gates. And though these gadgets cannot be presently just bought at

Apple stores or from any other electronics retailer, they can be designed in the lab and used by researchers for monitoring of signaling and regulatory small molecules directly in living cells.

Some of the computer/software giants took all these promising opportunities very seriously. For instance, a few years ago Microsoft formed the biological computation group with the tasks that include developing a new programming language for the takeaway DNA computing technology and designing molecular circuits made of DNA. The group has since developed a language called the DSD (for *DNA strand displacement*) and compiler (i.e., a program that converts instructions into a machine-code or lower-level form so that they can be read and executed by a computer), the basics of which you can now visually learn from YouTube at https://www.youtube.com/watch?v=_cEkIwMYypo. These new tools allow rapid prototyping and analysis in a convenient Web-based graphical interface of computational devices implemented using DNA, and they can also be used to design the DNA sequences needed to run DNA circuits [75, 76].

Also, to further promote research and developments in bio-logic gates, including the DNA-based ones, the online database was recently created that documents detailed information on nearly 200 experimentally validated logic gates of that kind, comprising AND (the most abundant in the database), OR, NOR, NOT, NAND, XOR, and others [77]. Given all this, the future looks quite optimistic for the use of DNA nanostructures in material science and technology. However, a number of real-life applications need the DNA self-assembly to substantially scale up both in quantities and in sizes, which is the theme of next section.

3.4 Towards Practical DNA Nanotechnology: The Race for Larger Quantities and Sizes, and Better Stability

Keeping up with the directions and applications of DNA is a never-ending job.
I. Eward Alcamo (American biologist), DNA Technology: The Awesome Skill (*2000*)

The DNA self-assemblies that were performed in numerous pilot studies described above required only small amounts of DNA, usually of the order of micrograms. But some proposed applications of structural DNA nanotechnology, such as the use of DNA nano-constructs for medical purposes or for building synthetic circuits, could only be practically realized if more DNA material will be used. One of the principal obstacles for scaling up the quantities of DNA nanostructures is the high cost of DNA oligonucleotides used as building blocks in DNA construction sets or as staples in DNA origami and obtained until recently through expensive solid-phase syntheses or enzymatic processes. To overcome this obstacle, the researchers from Technical University of Munich, one of Europe's leading universities, developed a new method for alternative making of short DNA strands, which should greatly reduce the cost of the hundreds of oligonucleotides and, therefore, may boost mass production of DNA nanostructures [78].

They employed bacterial viruses (aka bacteriophages) to produce long single-stranded artificial precursor DNA that contains hundreds of different oligonucleotide sequences. These sequences are separated by special DNA sequences with known self-cleaving activity—so-called DNAzymes [79]. A high-cell-density liter-scale cultivation of bacteriophage-infected *E. coli* bacteria in a stirred-tank bioreactor yielded gram-scale amounts of purified single-stranded DNA that subsequently cleaved itself into multiple oligonucleotides. Remarkably, this biotechnological production method reduces the cost of oligonucleotides by almost thousand times, as compared to chemical oligonucleotide synthesis.

Instead of employing bacteriophages and bacterial cells for price-efficient in vivo-based production of oligonucleotides, they can also be obtained in large quantities at low-cost in vitro in a cell-free process called rolling-circle amplification (RCA) [80]. In RCA reaction, the DNA polymerase enzyme generates a single-stranded DNA concatemer by moving around DNA minicircle and repeatedly synthesizing its linear replicas (see Fig. 3.23a). This robust enzymatic reaction takes place at constant temperature and inexpensively produces in test tubes long DNA templates with hundreds of repeated copies of the desired oligonucleotide sequences embedded within DNA minicircle, which can be cleaved out by other enzymes [81]. The group of Swedish researchers has recently used RCA to produce at once a pool of different oligonucleotides at predefined molar ratios (see Fig. 3.23b) and at a substantially lower cost than the price of their synthesis by chemical means—nearly 30 times cheaper [82]. Though this process is not as

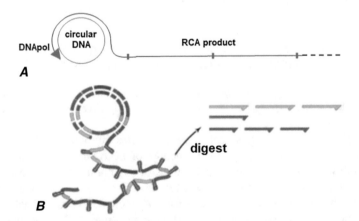

Fig. 3.23 Production of oligonucleotides by the RCA reaction. (**a**) Simplified schematics of the RCA process: DNA polymerase (DNApol; shown as arrowhead) multiply replicates circular DNA round by round to generate long single-stranded DNA molecules with tandem sequence repeats (aka concatemers; marked by vertical bars), which are the complements of circular DNA sequence. (**b**) Schematics of the RCA production of three different oligonucleotides shown by green, blue, and magenta colors at 3:1:3 copy ratios: circular DNA carries the corresponding numbers of complementary sequences of different oligonucleotides separated by short hairpin-forming sequences, which are used for enzymatic cleavage of the RCA product into desired oligonucleotides. (Adapted from [82])

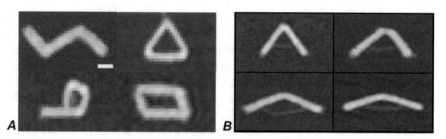

Fig. 3.24 DNA origami shape and hinge angle variation from the reference design. (**a**) Micrographs of different DNA nanostructures with multiple hinges assembled from the rodlike primary DNA origami construct (not shown here) with the reference set of 180 staples by replacing less than 13% of them with staples unique to each particular shape. Scale bar is 30 nm. (**b**) V-shaped DNA nanostructures with 60°, 105°, 120°, and 135° bent angles, respectively, obtained by varying the length of the adjuster strand connecting two arms of these constructs (seen as thinner lines in the middle of each shape) and replacing certain staples in the reference set. Only seven to ten new staples were required to introduce a wide range of angles into the rodlike primary DNA origami construct. (Adapted from [83])

economical as the in vivo-based one, it will still greatly reduce the price of DNA nanostructures and may also alleviate possible concern of involvement of live bacteria in manufacturing the DNA nano-constructs for medical applications (see the next section).

In parallel, Korean scientists suggested the original way to further reduce the cost of oligonucleotide staples in DNA origami [83]. They realized that a diverse variety of distinct DNA origami bent shapes could be designed with a single reference set of staples, majority of which will be shared by multiple different DNA origami nano-structures with controllable geometries and flexibilities. Then, only about one-tenth of them should be needed to be replaced by different staples necessary for each par-ticular DNA origami nanostructure. Such a staple-sharing approach can provide a versatile and cost-effective alternative in the design of DNA origami shapes with stiffness-tunable units (see Fig. 3.24), which could be used as convenient modules for hierarchical assemblies of larger DNA origami constructs (see below).

Besides the quantitative aspect discussed above in the first point, another key issue for real-world practical uses of DNA-based constructions is a matter of size: for future biomolecular engineering of synthetic materials that mimic dimensions and complexity of cells or subcellular components, assemblies of submicrometer- and micrometer-sized DNA structures would be necessary.[7] From DNA construc-tion sets, such as DNA tiles and DNA bricks, the DNA-based objects with micrometer sizes in one or two dimensions (like 1D and 2D DNA crystals and nanotubes) can readily be assembled (see Figs. 3.7 and 3.9) [17, 19, 22, 84]. Even the 250-μm-sized 3D DNA crystals have been obtained from the nanosized brick-like DNA building blocks [85], which could serve as a scaffold for constructing 3D

---

[7]To avoid possible confusion, this still will be the nanotechnology since all principal assemblies proceed at nanoscale from the nanosized DNA building blocks.

arrays of functional molecules or particles. But more complex 3D DNA objects of diverse shapes with all three sizes exceeding 100 nm can be obtained from DNA construction sets only at rather low yield and careful optimization of a self-assembly reaction is yet required for the production of DNA objects with reasonable yield when their sizes closely approach 100 nm [26, 86]. Hopefully, new DNA construction sets and/or new methods of assembly of DNA tiles and DNA bricks would be developed soon to go over 100 nm limit in 3D.

In DNA origami, the size of individual DNA nanostructures is limited by the length of the scaffold DNA strand, which is usually obtained from bacteriophage M13 being approximately 7000 nucleotides long (or 7 kb long), and can be folded into origami structures no more than 100 nm across. To go beyond this limit, the trio of Caltech scientists used square tiles of DNA origami as building units to hierarchically create 2D DNA origami arrays with nearly a micrometer across [87]. The square origami tiles join together through the formation of short DNA duplexes at their interfaces. For creating desired patterns on the array surface, origami tiles were decorated with DNA strands that extend from the origami surface at selected positions to serve as visual pixels on microscopy images (similar to the generation of electronic images on the TV screen). Several DNA "pictures" were drawn this way, including a chess board with figurines symbols and portrait of a woman resembling da Vinci's Mona Lisa (see Fig. 3.25a).

Another trio of scientists, from the same German research group that developed bacteriophage production of oligonucleotides described above, have made 3D DNA origami structures at sizes up to the micrometer scale by a different hierarchical self-assembly approach [88]. They used 50-nm-sized V-shaped DNA origami building blocks (called V-bricks), in which the angle of the V could be varied. By controlling the geometry and interactions between the building blocks,

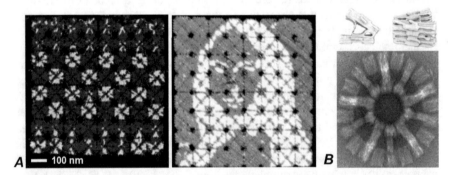

Fig. 3.25 Micrometer-scale 2D and 3D DNA origami constructs. (**a**) Micrographs of ~90 nm square origami tiles assembled into 8 × 8 tiles "chessboard" (left) or "portrait of a woman" (right). Visual pixels of chess figurines symbols and woman's face were created by DNA strands that extend from the origami surface at selected positions. (Adapted from [87].) (**b**) Micrograph of DNA dodecahedron with 450 nm diameter assembled from origami V-bricks, the side and the front views of which are shown above. (Obtained by microscopy; not in scale with the micrograph below; adapted from [88])

a large variety of higher order assemblies can be constructed. By using this method, the micrometer-long thick-wall tubes with 350 nm diameter, and polyhedrons up to 450 nm in diameter—the largest 3D objects obtained from DNA so far—were constructed of hundreds of DNA scaffolds and thousands of DNA staples (see Fig. 3.25b).

Alternative way to get DNA origami structures in the submicrometer and micrometer size ranges is to employ for their assembly longer DNA scaffold strands, other than those generally produced by bacteriophage M13, which could be generated by a variety of diverse approaches both in vivo and in vitro. For example, the team of academia researchers from North Carolina, USA, created hybrid bacteriophage from the two *E. coli* viruses, bacteriophages λ and M13, which replicates in special species of *E. coli* bacteria with single-stranded DNA more than 50 kb long [89]. Using these DNA single strands as origami scaffold, which is seven times longer than M13 DNA scaffold, both flat and non-flat 2D asymmetric origami sheets with shapes of notched rectangles and with controlled global curvature have been assembled. The largest sizes of these sheets were close to 300 nm and their surface area was over seven times larger than that obtained with M13 DNA scaffold.

Assembly of such large DNA origami sheets required over 1600 distinct DNA staple strands so that researchers developed new inexpensive process for DNA synthesis, which reduced the cost of staples by more than an order of magnitude. The thousand-plus pool of all necessary staple strands was made via an inkjet-printing process on a single chip embossed with functionalized micropillars made from cyclic olefin copolymer [89]. An alternative method for cost-reduced synthesis of oligonucleotides was a second achievement of this innovation in the DNA origami field.

A different, cell-free approach was utilized by the group of Chinese scientists who obtained in vitro a single-stranded scaffold DNA almost four times longer than M13 DNA by first producing the 26-kb-long double-stranded DNA with the well-developed molecular biotechnology polymerase chain reaction (PCR) followed by a selective enzymatic digestion of one of the two DNA strands [90]. Such a long DNA scaffold was folded by this research team into a 2D rectangular shape with the sizes of 238 × 108 nm using nearly 800 staple strands.

In 2009, the pioneer of DNA origami William Shih with two colleagues added a new twist on the DNA origami path to practice. Though before only the single-stranded scaffolds were used for DNA origami assemblies to avoid competitive reannealing with the complementary strand, Shih and co-workers have now demonstrated that successful assemblies could be achieved with a DNA scaffold provided initially in the double-stranded (ds) form [91].

To do this, they first heated the dsDNA in the presence of a chemical denaturing agent formamide to separate DNA strands, then added two distinct sets of staple strands, and quickly dropped the temperature while gradually removing the denaturant. Such a trick allowed the researchers to produce in a single folding event both 420-nm-long six-helix origami bundles and nearly 100-nm-sized nanostructures

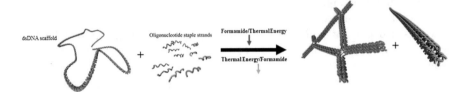

Fig. 3.26 Folding two separate DNA origami nanostructures from a dsDNA scaffold. High-temperature treatment combined with added formamide will result in the separation of dsDNA strands, which will become accessible to the two sets of staple strands. A sudden temperature drop followed by stepwise formamide removal will allow binding of staples to DNA single strands to initiate their folding into the two DNA origami nanostructures, while competitively preventing the reassociation of DNA single strands into initial dsDNA molecule. (Adapted from [91])

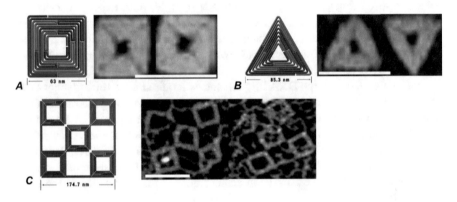

Fig. 3.27 Folding integrated DNA origami nanostructures from both strands of a dsDNA scaffold. (**a**) Square DNA nanostructure assembled with 2322-bp-long dsDNA. (**b**) Triangle DNA nanostructure assembled with 2027-bp-long dsDNA. (**c**) More complex five-unit square lattice DNA nanostructure assembled with 9280-bp-long dsDNA. In the schematics of nanostructures shown on the left the two scaffold DNA strands are drawn blue and red. Exemplary micrographs are shown on the right. All dsDNA scaffolds were obtained as fragments from enzymatic digestion of bacteriophage λ DNA. Scale bars are 100 nm. (Adapted from [92])

with intersecting six-helix origami bundles forming a core triangle and six short flaps extending from the triangle's vertices (Fig. 3.26).

A few years later, the Arizona State University (ASU) nanobioscientists, the same ASU research team that designed the above-described 3D DNA origami curved shapes [33] and DNA-based vaccine nanoparticles [42], have employed dsDNA molecules to fabricate integrated DNA origami structures that each incorporates both single-stranded DNAs of the initial DNA duplex as a unified scaffold. Accordingly, the two separated strands of dsDNA are aligned head to tail to effectively form single-stranded DNA scaffold twice as long as the individual single-DNA strands (see Fig. 3.27). In this unusual design, a reliable folding path should promote effective hybridization of correct staples while simultaneously suppressing the recovery of the initial dsDNA template.

To that end, in contrast to the origami assembly with dsDNA described above, the ASU team did not use any DNA denaturant, but only the optimized two cycles of heating and cooling in the presence of large excess of staple strands—first heating to 90 °C with rapid cooling to room temperature, and then heating to 45 °C with slow cooling to 4 °C [92]. Researchers believe that this new trick helps to kinetically trap both DNA strands in a metastable state of almost complete origami structures, while providing just enough thermal energy and time for the system to equilibrate to the final state.

By using this innovative approach, the ASU researchers assembled several DNA origami nanostructures, some of which are shown in Fig. 3.27. Since the sources of long single-stranded DNA are rather scarce, limiting the size and complexity of DNA nanostructures that can be assembled, it is anticipated that this alternative strategy will facilitate scaling up of DNA origami to greater complexity and mass production due to the virtually endless sources and the relative ease in obtaining natural and synthetic dsDNA molecules of larger lengths and diverse sequences, and also in relatively big quantities.

One more problem with DNA nanostructures that could limit their many potential applications is that they are in general rather labile constructions with could-be insufficient mechanical and thermal stability. Therefore, increasing the stability of DNA nano-constructs is also a key issue for their practical uses. Recently, the two research teams of German scientists have suggested two alternative solutions to this problem.

The Stuttgart University's team employed for this purpose disulfide cross-linking—nature's successful approach to stabilize folded proteins against structural disruptions. They assembled the DNA origami nanostructures with staple oligonucleotides bearing thiol groups that are readily capable of forming disulfide bonds by oxygen oxidation [93]. The staple-staple cross-linked DNA nano-constructs had significantly increased thermal stability, and they were converted back to non-cross-linked DNA nano-constructs under reducing conditions, which makes such a redox dependency an attractive feature for the construction of smart materials and biomedical applications.

Instead of chemical cross-linking of staple strands to enhance DNA origami stability, the team of researchers from Technical University Dresden and Paderborn University employed DNA ligase for enzymatically ligating all accessible staple strand ends directly without the need for synthetic nucleotide modifications [94]. This led to a considerable increase in overall stability of DNA origami nanostructures against their chemical and thermal denaturation.

With all these latest developments, it is clear that practical DNA nanotechnology is coming of age, and it is rapidly moving forward from the labs to commercialization. So, there is a hope that increasing size and complexity of DNA structures, and maximizing DNA production would finally lead researchers to the ability to build the DNA-based constructs for all imaginable real-world applications.

3.5   DNA Origami Constructs vs. DNA Construction Sets

If ever asked: What's more useful, the sun or the moon, respond: The moon. For the sun only shines during daytime, when it's light anyway, whereas the moon shines at night, when it is dark.

But, on the other hand, the Sun is better in that it shines and also makes us warm, whereas the Moon only shines; and that in the moonlit night only!

Kozma Prutkov, Fruits of Reflection: Thoughts and Aphorisms (*1853–1854*)

I expect that after reading the previous sections the reader may ask the following legitimate question: "What is the best way to assemble particular nanoshape from DNA—by using some DNA origami design or by using certain DNA construction set?" To answer this question, we need to consider a number of pros and cons for each approach because they both have their own advantages and limitations depending on the specific intended use of the projected DNA nanostructure.

For instance, for medical applications of DNA nanostructures those produced without the involvement of bacteria would be greatly preferable since even traces of dead bacteria bodies, if not properly removed from the final medication given to the patient, could trigger a life-threatening immune response called sepsis. It would therefore appear that DNA nanostructures assembled from the appropriate DNA construction sets could be more suitable for medicinal uses than those obtained by conventional DNA origami. Indeed, any DNA construction set can be cost-effectively prepared nonbiologically, and in therapeutically effective large amounts, e.g., by the above-described in vitro RCA production of DNA oligomers—efficient method that generates short DNA strands on gram scale and does not require bacterial culture [81]. And though both DNA origami staple strands and DNA scaffold can also be produced nonbiologically, e.g., by using the RCA method for making staple strands and the PCR technique for making scaffold strand, as it was described above [90], only biotech processes involving bacterial cultures would guarantee now the gram-scale production of scaffold DNA [95]. Moreover, while traditional DNA origami takes the scaffold strand from natural or genetically engineered bacteriophage DNAs, no biologically produced DNA is required in DNA brick assembly or any other DNA construction sets.

Medical-related concerns aside, the recent computational study, which compared ionic conductivity and mechanical properties of simulated DNA nanostructures assembled from DNA bricks with those produced by DNA origami with similar overall shapes, has concluded that DNA brick constructs were more leaky to ions, when placed in electrical field, than the shape-similar DNA origami counterparts [96]—essential property for prospective use of DNA assemblies in nanoelectronics (see Chap. 5). Yet, DNA origami constructs were found by this study to be more rigid when subject to external forces, as compared to equivalent DNA brick constructs. This important feature could shift the scales in the DNA origami in favor of certain nonmedical applications. Besides, as it was shown in the previous section, larger 3D nano-constructs can be obtained with hierarchically assembled DNA origami than with DNA bricks or DNA tiles, though the potential challenges with these origami superstructures could arise from the lack of sufficient firmness at connection points between the structural units [92].

There are two more points to comparatively consider here. I should notice that modular self-assembly of DNA nanostructures from DNA construction sets has generally lacked the complexity and exceptionally high design flexibility that DNA origami can offer. However, each distinct DNA origami shape would normally require a new scaffold routing design and the synthesis of a different set of staple strands. Such a poor reusability of staples has been one of the major hurdles to fabricate assorted DNA constructs by DNA origami in an effective way. In contrast, the modular assembly approach, when each component unit, such as DNA brick or DNA tile, can be included, excluded, or replaced locally and independently without altering the rest of the structure, offers a simpler approach to constructing various shapes, and also for fine-tuning the structure, if necessary.

In conclusion of this chapter, I would like to emphasize the evolution of bright ideas presented in it: from sculpting DNA shapes with simple DNA junctions and DNA crossovers to DNA origami and DNA bricks, and from simple geometric 2D and 3D DNA shapes to DNA-based viruslike vaccine nanoparticles, biosensors with DNA nanoantennas, and DNA molecular computing. All these developments suggest the existence of a vast design space that remains to be explored for the creation of novel DNA nanostructures and their innovative applications—we can only guess what is next.

Also note that in the material applications of DNA nanostructures described above, they were all used as essentially static objects, i.e., as molecular envelopes for encasing other molecules or as nanoscale frames and platforms to direct additional molecular assemblies. But researchers also realized that supplying energy to DNA nanostructures in the form of some molecular fuel could make them dynamic. That is how DNA machines and robots work and this will be the topic of the next chapter.

References

1. Seeman NC (1982) Nucleic acid junctions and lattices. J Theor Biol 99:237–247
2. Seeman NC, Kallenbach NR (1983) Design of immobile nucleic acid junctions. Biophys J 44:201–209
3. Nummelin S, Kommeri J, Kostiainen MA, Linko V (2018) Evolution of structural DNA nanotechnology. Adv Mater 30:e1703721
4. Ma RI et al (1986) Three-arm nucleic acid junctions are flexible. Nucleic Acids Res 14:9745–9753
5. Petrillo ML et al (1988) The ligation and flexibility of four-arm DNA junctions. Biopolymers 27:1337–1352
6. Shlyakhtenko LS et al (2000) Structure and dynamics of three-way DNA junctions: atomic force microscopy studies. Nucleic Acids Res 28:3472–3477
7. Wang YL, Mueller JE, Kemper B, Seeman NC (1991) Assembly and characterization of five-arm and six-arm DNA branched junctions. Biochemistry 30:5667–5674
8. Wang X, Seeman NC (2007) Assembly and characterization of 8-arm and 12-arm DNA branched junctions. J Am Chem Soc 129:8169–8176
9. Wang M, Afshan N, Kou B, Xiao SJ (2017) Self-assembly of DNA nanostructures using three-way junctions on small circular DNAs. ChemNanoMat 3:740–744

10. Han D et al (2013) DNA gridiron nanostructures based on four-arm junctions. Science 339:1412–1415
11. Chen JH, Seeman NC (1991) Synthesis from DNA of a molecule with the connectivity of a cube. Nature 350:631–633
12. Zhang Y, Seeman NC (1994) The construction of a DNA-truncated octahedron. J Am Chem Soc 116:1661–1669
13. Goodman RP et al (2005) Rapid chiral assembly of rigid DNA building blocks for molecular nanofabrication. Science 310:1661–1665
14. Kato T et al (2009) High-resolution structural analysis of a DNA nanostructure by cryoEM. Nano Lett 9:2747–2750
15. Aldaye FA, Sleiman HF (2007) Modular access to structurally switchable 3D discrete DNA assemblies. J Am Chem Soc 129:13376–13377
16. Li X, Yang X, Qi J, Seeman NC (1996) Antiparallel DNA double crossover molecules as components for nanoconstruction. J Am Chem Soc 118:6131–6140
17. Winfree E, Liu F, Wenzler LA, Seeman NC (1998) Design and self-assembly of two-dimensional DNA crystals. Nature 394:539–544
18. Sa-Ardyen P, Vologodskii AV, Seeman NC (2003) The flexibility of DNA double crossover molecules. Biophys J 84:3829–3837
19. Reishus D et al (2005) Self-assembly of DNA double-double crossover complexes into high-density, doubly connected, planar structures. J Am Chem Soc 127:17590–17591
20. Park SH et al (2005) Three-helix bundle DNA tiles self-assemble into 2D lattice or 1D templates for silver nanowires. Nano Lett 5:693–696
21. Liu D, Park SH, Reif JH, LaBean TH (2004) DNA nanotubes self-assembled from triple-crossover tiles as templates for conductive nanowires. Proc Natl Acad Sci U S A 101:717–722
22. Mathieu F et al (2005) Six-helix bundles designed from DNA. Nano Lett 5:661–665
23. Yin P et al (2008) Programming DNA tube circumferences. Science 321:824–826
24. Wei B, Dai M, Yin P (2012) Complex shapes self-assembled from single-stranded DNA tiles. Nature 485:623–626
25. Ke Y, Ong LL, Shih WM, Yin P (2012) Three-dimensional structures self-assembled from DNA bricks. Science 338:1177–1183
26. Ong LL et al (2017) Programmable self-assembly of three-dimensional nanostructures from 10,000 unique components. Nature 552:72–77
27. Shih WM, Quispe JD, Joyce GF (2004) A 1.7-kilobase single-stranded DNA that folds into a nanoscale octahedron. Nature 427:618–621
28. Rothemund PWK (2006) Folding DNA to create nanoscale shapes and patterns. Nature 440:297–302
29. Andersen ES et al (2008) DNA origami design of dolphin-shaped structures with flexible tails. ACS Nano 2:1213–1218
30. Andersen ES et al (2009) Self-assembly of a nanoscale DNA box with a controllable lid. Nature 459:73–76
31. Douglas SM et al (2009) Self-assembly of DNA into nanoscale three-dimensional shapes. Nature 459:414–418
32. Dietz H, Douglas SM, Shih WM (2009) Folding DNA into twisted and curved nanoscale shapes. Science 325:725–730
33. Han D et al (2011) DNA origami with complex curvatures in three-dimensional space. Science 332:342–346
34. Benson E et al (2015) DNA rendering of polyhedral meshes at the nanoscale. Nature 523:441–444
35. Keum JW, Bermudez H (2009) Enhanced resistance of DNA nanostructures to enzymatic digestion. Chem Commun 45:7036–7038
36. Linko V, Ora A, Kostiainen MA (2015) DNA nanostructures as smart drug-delivery vehicles and molecular devices. Trends Biotechnol 33:586–594

37. Okholm AH, Kjems J (2017) The utility of DNA nanostructures for drug delivery *in vivo*. Expert Opin Drug Deliv 14:137–139
38. Lee H et al (2012) Molecularly self-assembled nucleic acid nanoparticles for targeted *in vivo* siRNA delivery. Nat Nanotechnol 7:389–393
39. Huang Y, Huang W, Chan L, Zhou B, Chen T (2016) A multifunctional DNA origami as carrier of metal complexes to achieve enhanced tumoral delivery and nullified systemic toxicity. Biomaterials 103:183–196
40. Zhang C et al (2012) DNA-directed three-dimensional protein organization. Angew Chem Int Ed Engl 51:3382–3385
41. Ilyas M et al (2018) Atomic resolution structures of human bufaviruses determined by cryoelectron microscopy. Viruses 10:22
42. Liu X et al (2012) A DNA nanostructure platform for directed assembly of synthetic vaccines. Nano Lett 12:4254–4259
43. Veetil AT et al (2017) Cell-targetable DNA nanocapsules for spatiotemporal release of caged bioactive small molecules. Nat Nanotechnol 12:1183–1189
44. Bhatia D, Surana S, Chakraborty S, Koushika SP, Krishnan Y (2011) A synthetic icosahedral DNA-based host-cargo complex for functional in vivo imaging. Nat Commun 2:339
45. Torchilin V (2011) Tumor delivery of macromolecular drugs based on the EPR effect. Adv Drug Deliv Rev 63:131–135
46. Zhang Q et al (2014) DNA origami as an *in vivo* drug delivery vehicle for cancer therapy. ACS Nano 8:6633–6643
47. Jiang Q et al (2015) A self-assembled DNA origami-gold nanorod complex for cancer theranostics. Small 11:5134–5141
48. Amir Y et al (2014) Universal computing by DNA origami robots in a living animal. Nat Nanotechnol 9:353–357
49. Li Y et al (2004) Controlled assembly of dendrimer-like DNA. Nat Mater 3:38–42
50. Li Y, Cu YT, Luo D (2005) Multiplexed detection of pathogen DNA with DNA-based fluorescence nanobarcodes. Nat Biotechnol 23:885–889
51. Acuna GP et al (2012) Fluorescence enhancement at docking sites of DNA-directed self-assembled nanoantennas. Science 338:506–510
52. Shelbourne P et al (1993) Direct diagnosis of myotonic dystrophy with a disease-specific DNA marker. N Engl J Med 328:471–475
53. Lafontaine DA, Mercure S, Perreault JP (1998) Identification of a Crohn's disease specific transcript with potential as a diagnostic marker. Gut 42:878–882
54. Adleman LM (1994) Molecular computation of solutions to combinatorial problems. Science 266:1021–1024
55. Winfree E (1998). Algorithmic self-assembly of DNA. Ph.D. thesis, California Institute of Technology, Pasadena
56. Winfree E (2000) Algorithmic self-assembly of DNA: Theoretical motivations and 2D assembly experiments. J Biomol Struct Dyn 17(Suppl 1):263–270
57. Wang H (1960) Proving theorems by pattern recognition, I. Commun ACM 3:220–234
58. Wang H (1961) Proving theorems by pattern recognition, II. Bell System Tech J 40:1–41
59. Wang H (1965) Games, logic and computers. Sci Am 213(5):98–106
60. Grünbaum B, Shephard GC (1986) Tilings and patterns. Freeman, New York
61. Rothemund PWK (2000) Using lateral capillary forces to compute by self-assembly. Proc Natl Acad Sci U S A 97:984–989
62. Winfree E (2003) DNA computing by self-assembly. The Bridge 33(4):31–38
63. Amthor F (2014) Neurobiology for dummies. Wiley, Hoboken, NJ
64. Sahni V, Goswami D (2008) Nanocomputing: the future of computing. Tata McGraw-Hill, New Delhi
65. Currin A et al (2017) Computing exponentially faster: implementing a non-deterministic universal Turing machine using DNA. J R Soc Interface 14:20160990
66. de Silva AP, Uchiyama S (2007) Molecular logic and computing. Nat Nanotechnol 2:399–410

67. Mao C, LaBean T, Reif JH, Seeman NC (2000) Logical computation using algorithmic self-assembly of DNA triple-crossover molecules. Nature 407:493–496
68. Rothemund PW, Papadakis N, Winfree E (2004) Algorithmic self-assembly of DNA Sierpinski triangles. PLoS Biol 2:e424
69. Wang D et al (2014) Molecular logic gates on DNA origami nanostructures for microRNA diagnostics. Anal Chem 86:1932–1936
70. Divakaran V, Mann DL (2008) The emerging role of microRNAs in cardiac remodeling and heart failure. Circ Res 103:1072–1083
71. Benenson Y, Gil B, Ben-Dor U, Adar R, Shapiro E (2004) An autonomous molecular computer for logical control of gene expression. Nature 429:423–429
72. Kahan-Hanum M, Douek Y, Adar R, Shapiro E (2013) A library of programmable DNAzymes that operate in a cellular environment. Sci Rep 3:1535
73. Wei T et al (2017) Construction of DNA-based logic gates on nanostructured microelectrodes. Nucl Sci Tech 28:35
74. Seelig G, Soloveichik D, Zhang DY, Winfree E (2006) Enzyme-free nucleic acid logic circuits. Science 314:1585–1588
75. Lakin MR, Youssef S, Polo F, Emmott S, Phillips A (2011) Visual DSD: a design and analysis tool for DNA strand displacement systems. Bioinformatics 27:3211–3213
76. Lakin MR, Youssef S, Cardelli L, Phillips A (2012) Abstractions for DNA circuit design. J R Soc Interface 9:470–486
77. Wang L et al (2015) SynBioLGDB: a resource for experimentally validated logic gates in synthetic biology. Sci Rep 5:8090
78. Praetorius F et al (2017) Biotechnological mass production of DNA origami. Nature 552:84–87
79. Gu H, Furukawa K, Weinberg Z, Berenson DF, Breaker RR (2013) Small, highly active DNAs that hydrolyze DNA. J Am Chem Soc 135:9121–9129
80. Demidov VV (ed) (2016) Rolling circle amplification (RCA): towards new clinical diagnostics & therapeutics. Springer, Cham
81. Kool ET (1998) Rolling circle synthesis of oligonucleotides and amplification of select randomized circular oligonucleotides. US Patent 5,714,320
82. Ducani C, Kaul C, Moche M, Shih WM, Högberg B (2013) Enzymatic production of 'monoclonal stoichiometric' single-stranded DNA oligonucleotides. Nat Methods 10:647–652
83. Lee C, Lee JY, Kim DN (2017) Polymorphic design of DNA origami structures through mechanical control of modular components. Nat Commun 8:2067
84. Ke Y et al (2014) DNA brick crystals with prescribed depths. Nat Chem 6:994–1002
85. Zheng J et al (2009) From molecular to macroscopic *via* the rational design of a self-assembled 3D DNA crystal. Nature 461:74–77
86. Sajfutdinow M, Jacobs WM, Reinhardt A, Schneider C, Smith DM (2018) Direct observation and rational design of nucleation behavior in addressable self-assembly. Proc Natl Acad Sci U S A 115:E5877–E5886
87. Tikhomirov G, Petersen P, Qian L (2017) Fractal assembly of micrometre-scale DNA origami arrays with arbitrary patterns. Nature 552:67–71
88. Wagenbauer KF, Sigl C, Dietz H (2017) Gigadalton-scale shape-programmable DNA assemblies. Nature 552:78–83
89. Marchi AN, Saaem I, Vogen BN, Brown S, LaBean TH (2014) Toward larger DNA origami. Nano Lett 14:5740–5747
90. Zhang H et al (2012) Folding super-sized DNA origami with scaffold strands from long-range PCR. Chem Commun 48:6405–6407
91. Hogberg B, Liedl T, Shih WM (2009) Folding DNA origami from a double-stranded source of scaffold. J Am Chem Soc 131:9154–9155
92. Yang Y, Han D, Nangreave J, Liu Y, Yan H (2012) DNA origami with double-stranded DNA as a unified scaffold. ACS Nano 6:8209–8215
93. Wolfrum M et al (2019) Stabilizing DNA nanostructures through reversible disulfide cross-linking. Nanoscale 11:14921–14928

94. Ramakrishnan S et al (2019) Enhancing the stability of DNA origami nanostructures: staple strand redesign versus enzymatic ligation. Nanoscale 11:16270–16276
95. Kick B, Praetorius F, Dietz H, Weuster-Botz D (2015) Efficient production of single-stranded phage DNA as scaffolds for DNA origami. Nano Lett 15:4672–4676
96. Slone SM, Li CY, Yoo J, Aksimentiev A (2016) Molecular mechanics of DNA bricks: *in situ* structure, mechanical properties and ionic conductivity. New J Phys 18:055012

DNA Machines and Nanobots

4

> *Our hands, and the machines they operate, are simply too large to manipulate individual molecules. We must learn to program molecules to manipulate themselves.*
> David Doty *(American computer mathematician)*, Theory of Algorithmic Self-Assembly *(Communications of the ACM, 2012)*

The robust ability of DNA to self-assemble into miscellaneous nanostructures, as it was described in the previous chapter, can also be employed for the construction of molecular size active nanomechanical devices and nanorobots, aka molecular machines or nanomachines, made entirely or partially of DNA. This idea is so inspiring that in the past decade there was a burst of activity in the area of DNA machines: a diverse variety of them have been designed based on several different principles, and their workability has been proved using different physical techniques.

Please do not get confused: these are not miniaturized versions of common machines or mechanical devices like moving trucks or can openers—the term "DNA machine" is more generally applied to molecular DNA constructs that can switch between defined molecular conformations to mimic various mechanical functions that occur at the macroscopic level. Saying otherwise, DNA machines are DNA constructs that can do certain directed movements and motions, as well as some simple work, like picking up or moving nanoparticles and walking in controlled fashion [1]. Here, I present a few, the most smart of them in my opinion, due to their inventive design, which also will illustrate the major principles used so far in building the DNA machines.

V. V. Demidov, *DNA Beyond Genes*,
https://doi.org/10.1007/978-3-030-36434-2_4

4.1   Simple-Movement DNA Machines Controlled by Structural Transitions

The primary idea exploited in developing DNA machines was to use structural transitions between different DNA forms (aka secondary structures) as a driving force for moving the parts of a DNA-based molecular construct. Indeed, alterations in the DNA secondary structures induced by a certain external stimulus could cause reversible changes in the overall shape of a DNA construct resulting in kinematic spatial rearrangement of some of its parts, which is the basic principle of any mechanical device.

To the best of my knowledge, the very first nanomechanical device was of this kind and it was devised in the late 1990s by the pioneer of DNA nanotechnology Nadrian Seeman together with Alexander Vologodskii, recognized expert in physics of circular DNAs, and other co-workers from New York University (NYU) and University of Cologne (Germany). This DNA device is driven by changes in the length of arms of a four-way DNA junction within the supercoiled DNA molecule [2]. Supercoiling is a feature of circular DNA duplexes when they are twisted around each other, thereby forming a coiled DNA superhelix (or supercoil), as presented in Fig. 4.1.

Importantly due to the contorted and writhed nature of supercoiling, DNA supercoils are torsionally stressed, i.e., charged with energy, like a compressed torsion spring. And it is known that the degree of DNA supercoiling (and hence the level of torsional stress in a supercoil) can be controlled by special molecules, called intercalators, that invade the DNA double helix by inserting themselves between the stacks of DNA base pairs [3]. Based on this knowledge, the simple nanodevice was designed [2], which consists of a synthetic double-stranded circular supercoiled DNA molecule with incorporated partially mobile DNA branched junction (aka DNA cruciform). This molecule represents a system whereby torque applied to the circular molecule will have an impact on the cruciform, by relocating its branching point.

In operation of this device (see Fig. 4.2a), the addition and removal of DNA intercalating molecules (which could be simply controlled by light; see Sect. 4.2 below for photo-switching of azobenzene between intercalating and non-intercalating shapes) will change the twisting force inherent in supercoiled DNA. This will reversibly make the arms of a DNA cruciform shorter and longer, respectively, and therefore can bring closer or make more distant a pair of molecules attached to the ends of the cruciform arms. It is possible to choose such a pair of

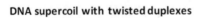

DNA supercoil with twisted duplexes   **untwisted (relaxed) dsDNA circle**

Fig. 4.1 Schematics of supercoiling in a circular DNA duplex. Relaxed, not supercoiled, dsDNA circle is also shown here for comparison

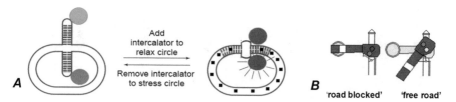

Fig. 4.2 DNA machine driven by intercalators. (**a**) This torsionally enforced system consists of a small double-stranded DNA circle (drawn by red lines), which is twisted like figure 8 (aka super-coiled, as in Fig. 4.1; not shown here for simplicity). DNA supercoil is designed to contain an extruded cross-shaped part, the cruciform, that partly relaxes the torsional stress caused by DNA supercoiling. The sequence of nucleotides of each stem of the cruciform allows them to form base pairs not only in the main body of a DNA circle but also in the extruded state, as shown schematically by blue strips. Addition of small molecules that intercalate into the double helix (shown at the right side as small black squares within the DNA duplex) further relaxes the torsional stress of DNA supercoiling, thus causing the cruciform to become shorter. A pair of molecules, which could signal the change of a shape of this construct, can be attached to the end of each cruciform arm (as shown by red and green circles). Indeed, a short cruciform brings a pair of signaling molecules closer to each other, which can also make one of them of a brighter color, and another one of a darker color due to the transfer of energy between them. (**b**) Mechanical semaphore used on some railroads to signal "Stop" and "Go" signs. You can see some similarity between the operation of this mechanism and the performance of DNA nanodevice shown and described in (**a**)

molecules as they will optically signal when close to each other using the so-called fluorescence resonance energy transfer (FRET) mechanism. This will create a kind of a semaphore switch similar to that used in the railroads (see Fig. 4.2b).

A set of different DNA nanomechanical devices with movable molecular arms were constructed by German researchers based on the knowledge that elevated concentrations of inorganic and organic multivalent cations, such as magnesium(2+), cobalt(3+), spermine(3+), and spermidine(4+), shield electrostatic repulsion between negatively charged DNA duplexes, which could bring them in close contact (Fig. 4.3a)—the process known as the DNA condensation [4]. One of these devices is shown in Fig. 4.3b and it consists of the two double-stranded DNA molecules bound to the three streptavidin proteins, with one of them flexibly connecting DNA duplexes into the V-shaped nano-construct.

By varying the concentrations of bivalent magnesium cations (Mg^{2+}) added to solution with this nano-construct, researchers found that it operates as the ion-dependent molecular switch: whereas at low Mg^{2+} concentrations this nanostructure adopts open conformation, higher Mg^{2+} concentrations (>5 mM) prompt the DNA duplexes to stick to each other, hence joining two distant streptavidin proteins together [5]. Magnesium-induced transition from open to close conformation is a reversible and fast process, and the DNA nanostructure can be readily attached to a surface by well-known techniques.

Thus I would like to speculate that as one of the potential applications of such nanodevice, it could be used as an ion-controllable valve in nanofluidic networks [6] to open or to plug certain fluidic nanochannels in response to the presence of various multivalent cations in flowing solutions, as it is shown schematically in Fig. 4.3c.

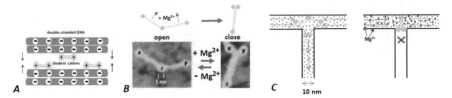

Fig. 4.3 DNA machine driven by magnesium salt. (**a**) Illustration of DNA condensation by biva-lent cations which negate electrostatic repulsion between two negatively charged DNA duplexes, bringing them in close contact. (**b**) Schematics and micrographs of magnesium-controllable V-shaped nano-construct composed of the two 50-nm-long, end-biotinylated DNA duplexes with attached protein molecules (streptavidin, marked as X, Y, and Z). Streptavidin is a globular tetra-meric protein with a diameter of 5 nm, which has particularly strong binding affinity for up to four biotin molecules. (Adapted from [5].) (**c**) Hypothetical operation of DNA nano-construct as shown in (**b**), as a nano-valve attached to nanofluidic T-junction. When magnesium ions are present in both flow-feeding nanochannels at a high concentration, they will cause the DNA nano-construct to adopt close conformation which will block the flow in T-junction. Evidently, other multivalent cations will control the flow in a similar way

And in case of magnesium, a simple sensor can be designed with this valve to detect elevated concentrations of magnesium in blood as a result of poor kidney functions (serious conditions known as hypermagnesemia), which at the levels above 5 mM cause life-threatening complications, such as low blood pressure, muscle paralysis, cardiac arrest, and coma [7].

Another simple-movement DNA machine driven by magnesium ions via induced structural transition was constructed by the Japanese research team who designed a DNA rotary motor based on the conformational differences between B-DNA and Z-DNA forms of the double helix [8]. Since Z-DNA is a left-handed double helix, as opposed to normal right-handed B-DNA form (see Glossary for description of DNA forms), Z-DNA-forming sequence experiences 180° rotation while changing its conformation from B- to Z-form. Therefore, under conditions favoring B–Z tran-sition DNA duplex with such a sequence will behave as a rotor, when placed between non-Z-DNA-forming duplexes.

The researchers integrated this rotor with a frame-shaped DNA origami sheet and also attached to it a flat paddle made of DNA origami, which allowed them to clearly observe in the micrographs the rotary movements in thus designed molecu-lar machine: the paddle rotates and moves forward when Z-DNA-promoting mag-nesium ions are added to solution in which DNA nano-construct is kept, and it rotates back when this solution is replaced with magnesium-free one to return to B-DNA form (see Fig. 4.4a, b). Alternatively, the DNA motor of that kind can be rotated forward and back just by thermocycling, since B-DNA reversibly transforms to Z-DNA at slightly elevated temperatures [9]. Also, the B–Z transition can readily be induced by DNA supercoiling, which is generated in DNA by a molecular size twister mechanism under the control of magnetic field [10].

Besides, the paddle blades attached to DNA rotor can be made by DNA origami in the form of a propeller (see Fig. 4.4c), which can create flow in nanofluidic net-works or drive DNA nano-construct forward like a motorboat, if a design enabling

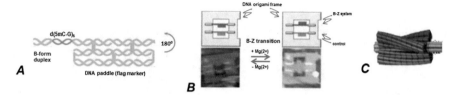

Fig. 4.4 Design and operation of a DNA rotary motor. (**a**) The motor comprises a duplex DNA "rod" connected to a DNA "paddle" via a short segment of Z-DNA-forming sequence (i.e., the stretch of methylated CGs). The paddle is created by DNA origami (see the previous chapter), and it serves as a flag marker for observing its rotation. (**b**) Micrographs of DNA rotary motor in action (bottom images), with top schematics explaining the details of its full assembly. Upon addition to the DNA motor-carrying solution of magnesium salt to promote B–Z transition, the DNA paddle in nano-motor rotates 180°, whereas no such movement can be observed in a control nano-device that misses B–Z rotor sequence (Adapted from [8]). (**c**) Computer-generated model of the DNA origami 40-nm-long nanorotor with three propeller blades, each consisting of six twisted DNA helix bundles. (Courtesy of Daniel Verschueren, Delft University of Technology)

permanent rotation of DNA rotor would be invented. Or it could also be used as a nano-turbine rotated by a flow in nanocapillaries or nanopores, as it was recently proposed by scientists from Delft University of Technology who are currently working towards this development [11].

4.2   Simple-Movement DNA Machines Driven by Hybridization

The above examples describe DNA nanomechanical devices driven by DNA structural transitions in response to addition-removal of molecules other than DNA. The first DNA machine driven and powered by DNA itself was reported in 2000 by research team from telecom equipment company Lucent Technologies, who constructed it based on DNA hybridization [12]. This nanomachine contains three DNA strands, A, B, and C, and it works as DNA tweezers (Fig. 4.5a). Strand A latches onto half of strand B and half of strand C joining them altogether, also acting as a hinge so that the two "arms," AB and AC, can move.

Initially, this DNA nanostructure has its arms open wide. The arms can be next pulled shut by adding a fourth strand of DNA (F; for "fuel") designed to stick to both of the dangling, unpaired sections of strands B and C, thus closing tweezers (Fig. 4.5b). The DNA strand F acts as a fuel for this molecular machine that is consumed to supply the energy for tweezers' motion provided by DNA base-pairing.

To reopen tweezers, one should just add an additional strand $\overline{F}$ having nucleotide sequence complementary to F to pair up with this strand by displacing strands B and C. Once F and $\overline{F}$ are paired up, they have no connection to the machine ABC, so they float away. The DNA machine can be opened and closed repeatedly by cycling between additions of strands F and $\overline{F}$. The total driving force for this nanomachine is the free energy released by hybridization of two DNA strands F and $\overline{F}$

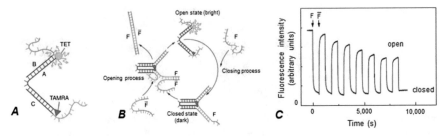

Fig. 4.5 Design and operation of a nanoscale DNA tweezers fueled by DNA. (**a**) This DNA machine is formed by hybridization of oligonucleotide strands A, B, and C. TET and TAMRA are a pair of detector molecules that optically indicate the state of the device via FRET effect, which makes one of them brighter or darker depending on their separation: when TET and TAMRA are well separated, TET brightly fluoresces green while exposed to light, whereas TET fluorescence is quenched in close proximity to TAMRA. (**b**) Schematics of closing and opening the molecular tweezers. "Closure" strand F hybridizes with the dangling ends of strands B and C (shown in blue and green) to pull the tweezers closed. Hybridization of "opener" strand $\overline{F}$ with the overhang section of F (red) allows $\overline{F}$ to remove F from the tweezers by strand displacement, forming a double-stranded "waste" product $F\overline{F}$ and allowing the tweezers to open again. (**c**) Real-time detection of multiple cycles of molecular tweezers' closing and opening by following TET fluorescence after sequential addition of strands F and $\overline{F}$. (Adapted from [12])

while producing a waste molecule $F\overline{F}$ and can be converted into mechanical movement of loads when the nanomachine is loaded.

The tweezers' closing and opening have been proven by tagging strand A at either end with light-absorbing/emitting molecules so that they shine bright when tweezers are open, whereas they do not emit light when they are adjacent to each other in the tweezers' closed state due to the so-called quenching effect. Such tweezers can be used for moving small molecules from one place to another, for instance, when sorting out molecules type I from molecules type II, or, if combined with a molecular cage, to alternately hide and reveal a target molecular group.

Two years later, a duo of American bio/nano-researchers from the University of Florida made another hybridization-driven DNA nanomachine which operates as a nanoscale switch by swapping itself between two DNA conformations, intramolecular quadruplex (see Glossary) and intermolecular duplex [13]. The working principle of this nanomachine is shown in Fig. 4.6a: it performs the cycles of extending-shrinking motion through alternating DNA hybridization and strand displacement reactions caused by sequential additions of complementary fuel strands α and β. Remarkably, these dynamic cycles are somewhat reminiscent of characteristic movements of certain small invertebrate animals—hydras, leeches, and some caterpillars (see Fig. 4.6b), as they make their way ahead in a looping/arching fashion [14–16]. This kind of periodic motions also inspired the design of a number of smart small robots which move forward by cycles of arching movements and transfer with them some cargo load [17, 18].

The nanomachine is simply a short guanine-rich DNA single strand able to fold into an intramolecular quadruplex (or tetraplex) compact structure, in which its two

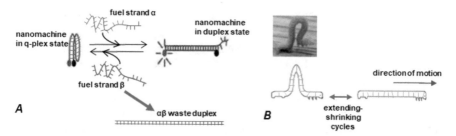

Fig. 4.6 Design and operation of a nanoscale DNA switch fueled by DNA. (**a**) 17-mer DNA single strand with G-rich nucleotide sequence $TG_2T_2G_2TGTG_2T_2G_2T$ can adopt by itself an intramolecular quadruplex conformation, as it is shown on the left. It extends into duplex when it hybridizes to strand α and shrinks back to quadruplex when strand α is displaced via hybridization with strand β (DNA strands α and β are totally complementary to each other). The blue portion in α is complementary to the DNA nanomachine, and the purple portion represents extra bases to form overhanging sticky ends for subsequently binding β in nanomachine's duplex state. As in case of DNA tweezer described above, tagging G-rich strand at both ends with light-absorbing/emitting molecules allows to follow the periodic movements of this nanoscale DNA switch: they shine bright when nanomachine is extended, whereas they are dark in a shrunken state. (**b**) This sketch illustrates the similarities found by the nanomachine designers between extending-shrinking motions of a nanoscale DNA switch and arching movements of some caterpillars, which make their journey by move-the-front-then-move-the-back process. (Photo of caterpillar in the looped shape is from iStockphoto/Eric Shaw)

ends are kept close to each other. In the presence of fuel strand α, which has a sequence complementary to the guanine-rich DNA strand, these two DNA strands form intramolecular duplex with dangling end, thus straightening the quadruplex and separating the two ends away from each other. Addition of fuel strand β initiates the strand exchange reaction, in which short guanine-rich DNA strand is displaced by a longer strand β to form more stable duplex, waste $\alpha\beta$, and to return the nanomachine from extended state into shrunken state, thus finishing one extending-shrinking cycle. And adding of more α strands will start a new working cycle with its completion being achieved by adding of more β strands.

To easily operate this DNA nanodevice, the researchers placed it inside of a microchannel, where the solution of strands α and β periodically flew in and out, and they found that the device was acting as a nanoscale switch (similar to that shown in Fig. 4.3c) by continuously cycling extending-shrinking motions dozen times without losing its workability. Besides, this nanodevice can be used for manipulating nanoparticles since it could develop the force comparable to the forces exhibited by natural motor proteins, kinesins and myosins, that move diverse cargo molecules, vesicles, and organelles in live cells.

Another hybridization-driven tweezer-like DNA machine has been designed by the Purdue University scientists based on a DNA enzyme (aka DNAzyme), which is an artificial DNA construct capable of performing a catalytic chemical reaction that cleaves RNA (but not DNA) hybridized to it [19]. This molecular machine is fueled by RNA, the second natural nucleic acid chemically and genetically related to DNA (see Glossary for similarities and differences between RNA and DNA). Unlike the

DNA machines described above, operation of which needs to be controlled by some operator who must periodically add the necessary components or change the solution conditions, the DNA machine fueled by RNA is fully autonomous, self-operating nanomechanical device, not requiring any human intervention or control to perform its function.

The design and operation of this nanomachine are shown in Fig. 4.7: the triangular shaped DNA construct automatically swings back and forth between the two states, one of which can be called "closed," and another one "open." For functioning properly, the DNAzyme requires magnesium(2+) divalent cations, which cause its single DNA strand to collapse into a compact coil, if it is not bound to the RNA substrate, thus keeping this nanodevice in the closed state. To open, the device binds a strand of RNA that hybridizes to so far inactive DNAzyme strand. Next, the RNA is cleaved by an active DNAzyme formed on this side of the device. When the RNA is cleaved, the two cleavage products are sufficiently short so that they dissociate from the device; and when the products dissociate, the device closes.

Then another RNA molecule from the solution binds this DNA machine so that the system will go through another round of "close-to-open" motion, and so on and so forth. To stop the machine, a DNA "brake" strand that cannot be cleaved by the DNAzyme can be added to the solution. And the machine can be restarted again by removing the brake via strand displacement through competitively binding it to a complementary DNA strand initiated at break's overhang (see Fig. 4.7).

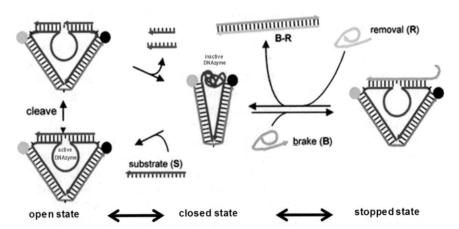

Fig. 4.7 Design and operation of DNA tweezers fueled by RNA. This DNA machine comprises a DNAzyme to bind and to cleave a piece of RNA fuel. When RNA binds, the machine goes to the open state. Following cleavage of the RNA, the products are sufficiently short that they dissociate and the device closes. However, another RNA strand in the solution can bind and restore the machine to the open state. The state of the machine can be optically monitored by FRET using the dyes represented by black and green circles. As an option, a "brake" can be removably applied to the machine: a strand of DNA (green) can block the RNA-binding site, but can be removed by a complementary DNA strand (light blue). (Adapted from [19])

4.3   DNA Walkers and DNA Pickers

A different autonomous DNA nanomachine also driven by hybridization but capable of more complicated movements was ingeniously designed by Andrew Turberfield and Nadrian Seeman with co-workers who independently created two variants of a bipedal DNA walker that uses two single-stranded DNA "legs" for self-moving forward [20, 21]. DNA walker travels step by step along specially designed DNA track containing footholds that provide the anchoring sites for the attachment of walker legs, and its movement is driven by hybridization of DNA fuel strands that detach walker legs from footholds one step at a time (Fig. 4.8). This DNA machine should be able to fulfill roles that entail the performance of useful mechanical work on the nanometer scale, e.g., to serve as a molecular carrier in the robotic nano-devices.

I should note here that the use of DNA or RNA as a fuel or brake for DNA machines of the kind described above generates the "waste" nucleic acid duplexes in every working cycle of DNA machine, which will be gradually accumulated during the machine's continuous operation. The accrued "wastes" will inevitably impede efficient operation of DNA machine producing them so that such "wastes" need to be somehow removed or cleaned up from time to time, which would require certain additional manipulations. To overcome this apparent complication, the DNA machines powered by light were independently developed by Japanese and American researchers [22, 23]. The employment of light as a driven force also greatly simplifies the design and operation of DNA machines.

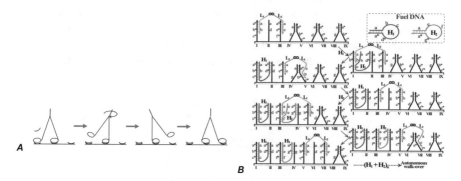

Fig. 4.8 Autonomous bipedal DNA walker. (**a**) Cartoon schematically showing the sequence of movements of DNA walker during a single footstep driven by competitive hybridization of a fuel DNA strand (red stroke line) to a foothold. (**b**) More detailed depiction of operation of single-stranded DNA walker comprising legs L1 and L2, which moves over a ladder-like DNA track with footholds I to IX by using two DNA hairpins H1 and H2 as fuels. Initially, the walker's legs are bound by hybridization to a pair of footholds I and III. Next, a first fuel hairpin displaces one of the legs, thus releasing it for binding to one of the neighboring foothold. Then the same process is repeated with another leg, as driving by a second fuel hairpin. As a result, the walker moves ahead along the DNA track by the two full steps. Cyclic repetition of this process results in a continuous translocation of DNA walker down the DNA track. (Reproduced with permission from ref. [1])

The light-driven DNA machines comprise DNA strands with incorporated azobenzene dye moieties, which can be in either planar *trans* or folded *cis* isomeric configurations. When azobenzene is irradiated with ultraviolet light, it quickly isomerizes from the *trans* form to the *cis* form. On the contrary, irradiation of *cis*-azobenzene with visible light returns it back to the *trans* form. This makes the azobenzene-coupled DNA nanomachines to be responsive to light: the planar *trans*-azobenzene intercalates between adjacent DNA base pairs and stabilizes DNA duplex by stacking interactions, whereas the nonplanar *cis*-azobenzene destabilizes it by steric hindrance. Therefore, hybridization of such photo-responsive DNAs can be efficiently turned "on" and "off" by light pulses of different wavelengths so that they can be utilized as reusable fuel strands for switching DNA machines between two different states (usually their open and closed conformations) without creating any "wastes."

DNA machines can also be controlled by protein binding: Fig. 4.9 shows schematically the performance of a modular antibody-powered DNA-based nanomachine that has recently been designed by the joint team of Italian and Canadian researchers [24]. This DNA machine could be called as a DNA picker, and can load and release a molecular cargo to be used, for example, in drug delivery systems for controlled drug release.

Researchers from California Institute of Technology (Caltech) have made lately the next great move: they combined DNA walker and DNA picker in a smart DNA robotic device that is capable of performing both cargo-picking and cargo-sorting tasks [25, 26]. The two-feet, single-stranded DNA walker was upgraded with one hand, and it walked along a flat surface built by DNA origami to pick up different cargo molecules distributed at certain location, and then to selectively drop them off at destinations specified for each cargo type (Fig. 4.10).

This DNA nanorobotic machine makes nearly 300 steps while sorting the cargos, and it takes more than 10 h for one DNA robot to complete the cargo sorting task. Amazingly, this study shows that to speed up the cargo sorting, several DNA robots can work cooperatively and four such robots will accomplish the task in less than

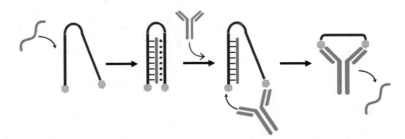

Fig. 4.9 Antibody-powered tweezers-like DNA-based nanomachine. This DNA machine is able to load itself with a cargo DNA strand and then release it upon binding to a specific antibody. The core of this machine is a DNA strand (black) labeled at its termini with two antigens (green hexagons). This DNA strand can bind another DNA strand (blue) by forming a clamp-like DNA triplex. The binding of an antibody (shown here as a Y-shaped molecule) to the two antigens opens the cargo-loaded clamp, which disrupts the triple-DNA complex with the consequent release of the loaded DNA strand from unstable DNA duplex. (Adapted from [24])

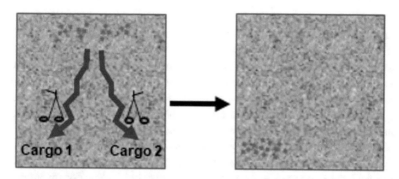

Fig. 4.10 Schematics of DNA nanorobotic sorter at work. Left: Both type 1 and type 2 cargo molecules, shown as blue and green bright spots, were distributed at their initial location near the top of a flat DNA origami surface. This surface has footholds for DNA walker and the two "storage" areas at the bottom left and right corners, each being destined to anchor a particular cargo type. Right: After about 15 h, the single-handed DNA nanobot randomly walking on the surface picked up and moved cargo molecules of the first type to the intended location at a left bottom corner and cargo molecules of the second type to the intended location at a right bottom corner. (See [25] for actual micrographs)

1 h! And Caltech researchers hope that with simple communication between the DNA walking robots, they could perform even more sophisticated tasks so that these molecular machines could eventually be easily programmed like macroscopic robots, but working in microscopic environments.

4.4   Remotely Controlled DNA Machines

Evidently, the effective use of dynamic DNA nanostructures as nanorobots requires fast and reliable mechanisms for their actuation. Robust design of DNA machines powered by light described above is one of the possible solutions. One more exciting development in DNA nanorobotics has recently been made by a team of Israeli scientists who demonstrated that opening of closed DNA nano-constructs can be done by precise local heating and that this process can be activated from the distance [27]. The team employed the barrel-shaped DNA origami nanocapsules designed by them before (as described in the previous chapter) that were closed by a gate locked by DNA strands hybridized to each other. But instead of using proteins as keys that open the gate, as it was done previously, this time the gate was open, i.e., literally melted, by the heat from metallic nanoparticles attached to it. Thus devised DNA nanocapsules were loaded with drugs and closed by a locked gate, and then they were injected into live cockroaches. Next, insects were subjected to radio waves, which heated nanoparticles by electromagnetic energy and opened the lock, exposing therefore the drug to the insect's body.

Though the DNA nano-constructs employed in this study are not exactly the nanomechanical robotic devices since they cannot perform any regulated movements or dynamic work, the study has principally proved that certain actions of DNA nanorobots can be remotely controlled by radio waves in a sense like macro-scale mechanical robots are normally controlled [28]. What is more, Israeli researchers linked the radio-wave emitter with computer which detected specific changes in the electrical activity of human brain and turned on the emitter, accordingly [27]. These researchers believe that novel technology enables the online switching of a drug molecule on and off by brainwaves in response to a person's mental or cognitive state. This may give a new method of therapeutic control in psychiatric disorders such as schizophrenia, depression, attention deficits, and hyperactivity, which are among the most challenging conditions to diagnose and treat.

And of course, the remotely controlled DNA nanorobots can be used for delicate precise microsurgery and nanosurgery of the future, when a surgeon will be able to perform noninvasive surgery at the cellular level [29]. For instance, these surgical nanorobots would find and eliminate micrometastases or even isolated cancerous cells, remove microvascular obstructions, and conduct molecular repairs on traumatized extracellular structures. To perform these tasks, DNA nanorobots will be equipped with the ability to move inside the body by using semiautonomous DNA motors and to search for pathology by specific biomolecular interactions, and then to correct it by nanomanipulation, e.g., via nano-assisted radiofrequency thermal ablation—all this is possible based on the developments described above and the like. Alternatively, surgical nanorobots could be built from the non-DNA nano-modules to possess full array of autonomous and remotely controlled subsystems, including onboard sensors, motors, manipulators, power supplies, and molecular computers, but getting all these nanoscale components to self-assemble in the right sequence can rationally be done only by using sticky DNA handles attached to them.

That kind of surgery was imagined more than 50 years ago in the original science fiction movie "Fantastic Voyage" directed by Richard Fleischer and written by Harry Kleiner, and in its adaptation novel by Isaac Asimov with the same name [30]. This fiction is about a medical crew aboard a submarine which were altogether reduced to microscopic sizes by innovative secret technology that can miniaturize matter by shrinking individual atoms. The submarine, which became now of the size of a microbe, was then injected into the body of a patient in order to find and to destroy a deadly blood clot in his brain that no surgery can remove from the outside. Although the crew faced many obstacles during the mission (and also afterwards), the clot was finally reached and successfully removed.

Externally applied electrical or magnetic fields to actuate and dynamically manipulate the DNA robots are another robust option for their control without physical contact. Indeed, the robustness of using electrical fields for such purposes has been demonstrated recently by the team of German nanotech researchers. They built the nanoscale manipulator with the DNA-made robotic arm controlled by electric field, which can be used for electrically driven fast and precise transport of molecules or nanoparticles over tens of nanometers [31]. Since DNA is a negatively charged macromolecule, the DNA-made robotic arm will be forced to move by dynamic electric fields in a desired way. Figure 4.11 shows this device in rotatable operation when DNA nanorobotic arm completes the whole round rotation in less than a second by switching of the arm between different positions on the platform within milliseconds

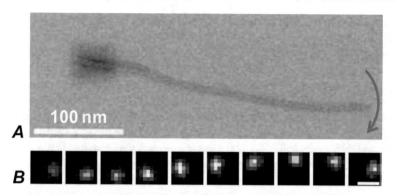

Fig. 4.11 DNA-based molecular platform with an integrated robotic arm. (**a**) Micrograph of the device composed of the rectangular DNA origami plate with an integrated rotatable (as indicated by blue arrow), more than 300-nm-long robotic arm made of a rigid DNA six-helix bundle. (**b**) A series of fluorescent micrographs demonstrating rotation of the DNA-made robotic arm by the action of electric field applied externally. All shots were done in about 100 ms one after another. The static end of the DNA arm is positioned at the center of these images, whereas the freely movable end of this arm is located near the edges of these images, and it is labeled with fluorophores to easily observe its rotation. Scale bar is 1 μm. (Courtesy of Friedrich Simmel, Technical University Munich)

by a rotating electrical field. The ability of DNA robotic arm for electrically controlled movement of cargoes was proved by transporting attached fluorescent molecules between two docking sites on the DNA origami platform, and by attaching gold nanoparticle to one side of the arm and rotating it with different speed.

A few months later, multidisciplinary research team from the Ohio State University has reported that externally applied magnetic fields could also be used for actuation and manipulation of the DNA-made robotic arms with sub-second response times [32]. They fixed magnetic microbead at one end of micron-long rod-like lever assembled by DNA origami and rotatably attached at another end to a surface, and continuously rotated this DNA nanorobotic arm by a rotating magnetic field with angular frequency up to two rotations per second, similar to electrically driven rotations of DNA-made arm described above. Moreover, the researchers showed that two DNA lever arms could be magnetically rotated at the same time, and even opposite to each other if magnetic moment of the beads attached to them were orientated in different directions. They are also confident that more complex magnetic manipulation platforms could be used to simultaneously operate multiple DNA nanomachines performing different movements in parallel.

Combined with appropriate pickup and release mechanisms, some of which have already been successfully tested by others as described above, it is conceivable that these two innovative technologies for remote manipulation of DNA nanorobots could be an important step towards the realization of nanoscale chemical synthesis factories. So far merely the stuff of science fantasy, these nanofactories may become one day the last word in chemical manufacturing [33]. Just imagine that a large number of nanorobotic arms built of DNA perform in parallel electrically or magnetically coordinated synthesis of various desired molecules by picking up different reactive molecules and bringing them together in required pairs for their interaction, like it happens within live cells during the protein biosynthesis.

For this vital cellular process, protein synthesis factories called polysomes (or polyribosomes) are self-assembled within a cell, which are complexes of individual messenger RNA molecules (mRNAs) with attached numerous ribosomes—Nature's multicomponent molecular machines that translate mRNA code into chains of protein polypeptides (see Fig. 4.12a, b). Multiple polysomes form one after another on genomic DNA as several RNA polymerase enzymes simultaneously read the DNA gene and successively synthesize a number of identical mRNA molecules in a row (Fig. 4.12c). In polysomes, a bunch of ribosomes move along mRNA molecule encoding particular protein, and they work in parallel by picking independently one particular amino acid after another one from a pool of almost two dozen different free amino acids located nearby at very low concentrations and then attaching them one by one to the growing protein chains.

Such a biosynthetic approach to generating new molecules is fundamentally different from conventional chemical synthesis when for each synthetic step chemists combine in one solution a specific set of only a few reactants at high concentrations, to discard them thereafter before the next synthetic step with another set of a few high-concentration reactants [34]. And it is evidently much more efficient because of high synthesis speed and specificities that cannot be achieved with conventional synthetic methods. Therefore, the anticipated parallel-running chemical synthesis with nanoscale DNA-based nanorobotic factories would essentially mimic Nature's strategy for effectively controlling chemical reactivity applied to synthetic molecules.

One more significant step towards this fantastic goal was done a few years ago by the Ohio State University team who proved that DNA origami design makes it possible to engineer complex and reversible motions of nanoscale elements of DNA machines and robots, which functionally mimic the construction and operation of

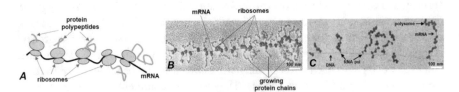

Fig. 4.12 Cellular protein synthesis factories. (**a**) Schematics of polysome, which is a large group of ribosomes attached to one mRNA strand (only a few are drawn there for simplicity). Ribosomes synthesize protein polypeptides by linking dozens of amino acids together in the order specified by mRNA molecules. A single ribosome is composed of small and large subunits, with the small subunit reading the mRNA code and the large subunit joining the corresponding amino acids to form a polypeptide chain. Each polypeptide chain gets longer as ribosomes move forward along mRNA to finally produce the complete protein. (**b**) Micrograph of an individual polysome with clearly visible growing protein polypeptide chains. (Adapted from [35]). (**c**) Micrograph showing a number of polysomes that grow successively on a chromosome from bacterium *E. coli*. (Courtesy of late Charles A. Thomas, Jr., Harvard Medical School)

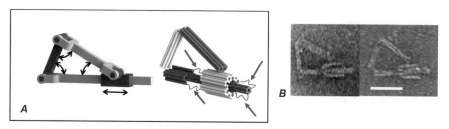

Fig. 4.13 DNA origami design of a slider-crank mechanism. (**a**) Macroscale solid model of this mechanism (left) and molecular scale model of its DNA origami counterpart (right). The slider-crank has three revolute joints and one sliding joint, as indicated by black arrows. In the DNA origami design, cylinders represent DNA double helices; the zigzag lines indicated by red arrows represent flexible stretches of single-stranded DNA, which connect the DNA origami parts and allow their relative motion. (**b**) Micrograph of the two representative slider-cranks assembled by DNA origami. Scale bar is 50 nm. (Courtesy of Carlos Castro, the Ohio State University)

elaborate macroscopic mechanical devices and kinematic mechanisms [36]. Figure 4.13 illustrates their approach by showing the design of a nanoscale slider-crank linkage—important module of many robotic mechanisms, including robotic legs and arms—which was successfully constructed by DNA origami in a totally functional state that coupled rotational and linear motions featured by its macroscopic counterpart.

Besides several DNA machines and nanorobots presented in this chapter, a large variety of other DNA-based machines have been constructed during the past few years comprising different DNA structural motives as construction elements, such as DNA quadruplexes, aptamers, catenanes, rotaxanes, and more, aiming to enrich the toolbox of DNA machines for miscellaneous purposes. I therefore refer the interested reader to the recent comprehensive reviews, which meticulously describe and nicely portray several dozens of these diverse nano-devices and their prospective applications [1, 37, 38]. And although substantial progress in the development of DNA machines has already been accomplished, researchers are still learning more and more new things in building synthetic molecular machinery from DNA. Their practical applications are yet in progress and multidisciplinary efforts would most likely be necessary to finally bring DNA machines to real life.

References

1. Wang F, Willner B, Willner I (2014) DNA-based machines. Top Curr Chem 354:279–338
2. Yang X, Vologodskii AV, Liu B, Kemper B, Seeman NC (1998) Torsional control of double-stranded DNA branch migration. Biopolymers 45:69–83
3. Vologodskii A (2019) Topology and physics of circular DNA. CRC Press, Boca Raton
4. Koltover I, Wagner K, Safinya CR (2000) DNA condensation in two dimensions. Proc Natl Acad Sci U S A 97:14046–14051
5. Niemeyer CM et al (2001) Nucleic acid supercoiling as a means for ionic switching of DNA-nanoparticle networks. ChemBioChem 2:260–264
6. Edel J, Kim MJ, Ivanov A (eds) (2016) Nanofluidics. Royal Society of Chemistry, Cambridge

7. Jahnen-Dechent W, Ketteler M (2012) Magnesium basics. Clin Kidney J 5(Suppl 1):i3–i14
8. Rajendran A, Endo M, Hidaka K, Sugiyama H (2013) Direct and real-time observation of rotary movement of a DNA nanomechanical device. J Am Chem Soc 135:1117–1123
9. Chaires JB, Sturtevant JM (1986) Thermodynamics of the B to Z transition in poly(m5dG-dC). Proc Natl Acad Sci U S A 83:5479–5483
10. Lee M, Kim SH, Hong SC (2010) Minute negative superhelicity is sufficient to induce the B–Z transition in the presence of low tension. Proc Natl Acad Sci U S A 107:4985–4990
11. Verschueren D (2018) Plasmonic Nanopores for Single Molecule Sensing. Chapter 10. Towards flow-driven rotation of a DNA origami nanomotor. Doctorate thesis, Delft University of Technology. https://doi.org/10.4233/uuid:a0099c3d-3244-4789-baa7-4819d7a429fa
12. Yurke B, Turberfield AJ, Mills AP Jr, Simmel FC, Neumann JL (2000) A DNA-fuelled molecular machine made of DNA. Nature 406:605–608
13. Li JJ, Tan W (2002) A single DNA molecule nanomotor. Nano Lett 2:315–318
14. Passano LM, McCullough CB (1964) Co-ordinating systems and behaviour in *Hydra*: I. Pacemaker system of the periodic contractions. J Exp Biol 41:643–664
15. Harley CM, Rossi M, Cienfuegos J, Wagenaar D (2013) Discontinuous locomotion and prey sensing in the leech. J Exp Biol 216:1890–1897
16. Plaut RH (2015) Mathematical model of inchworm locomotion. Int J Non-Lin Mech 76:56–63
17. Moreira F, Abundis A, Aguirre M, Castillo J, Bhounsule P (2018) An inchworm-inspired robot based on modular body, electronics and passive friction pads performing the two-anchor crawl gait. J Bionic Eng 15:820–826
18. Ning J, Ti C, Liu Y (2017) Inchworm inspired pneumatic soft robot based on friction hysteresis. J Robot Autom 1:54–63
19. Chen Y, Mao C (2004) Putting a brake on an autonomous DNA nanomotor. J Am Chem Soc 126:8626–8627
20. Green SJ, Bath J, Turberfield AJ (2008) Coordinated chemomechanical cycles: a mechanism for autonomous molecular motion. Phys Rev Lett 101:238101
21. Omabegho T, Sha R, Seeman NC (2009) A bipedal DNA Brownian motor with coordinated legs. Science 324(5923):67–71
22. Liang X, Nishioka H, Takenaka N, Asanuma H (2008) A DNA nanomachine powered by light irradiation. ChemBioChem 9:702–705
23. Kang H et al (2009) Single-DNA molecule nanomotor regulated by photons. Nano Lett 9:2690–2696
24. Ranallo S, Prevost-Tremblay C, Idili A, Vallee-Belisle A, Ricci F (2017) Antibody-powered nucleic acid release using a DNA-based nanomachine. Nat Commun 8:15150
25. Thubagere AJ et al (2017) A cargo-sorting DNA robot. Science 357:eaan6558
26. Reif JH (2017) DNA robots sort as they walk. Science 357(6356):1095–1096
27. Arnon S et al (2016) Thought-controlled nanoscale robots in a living host. PLoS One 11:e0161227
28. Deal WFIII, Hsiung SC (2007) Exploring telerobotics: a radio-controlled robot. Technol Teach 10:11–17
29. Singh S, Singh A (2013) Current status of nanomedicine and nanosurgery. Anesth Essays Res 7:237–242
30. Asimov I (1966) Fantastic voyage. Houghton Mifflin, Boston
31. Kopperger E et al (2018) A self-assembled nanoscale robotic arm controlled by electric fields. Science 359:296–301
32. Lauback S et al (2018) Real-time magnetic actuation of DNA nanodevices via modular integration with stiff micro-levers. Nat Commun 9:1446
33. Simmel FC (2012) DNA-based assembly lines and nanofactories. Curr Opin Biotechnol 23:516–521
34. Li X, Liu DR (2004) DNA-templated organic synthesis: nature's strategy for controlling chemical reactivity applied to synthetic molecules. Angew Chem Int Ed Engl 43:4848–4870

35. Francke C, Edstrom JE, McDowall AW, Miller OL Jr (1982) Electron microscopic visualization of a discrete class of giant translation units in salivary gland cells of *Chironomus tentans*. EMBO J 1:59–62
36. Marras AE, Zhou L, Su IIJ, Castro CE (2015) Programmable motion of DNA origami mechanisms. Proc Natl Acad Sci U S A 112:713–718
37. Chandran H, Gopalkrishnan N, Reif J (2013) DNA nanorobotics. In: Mavroidis C, Ferreira A (eds) Nanorobotics. Springer, New York, pp 355–382
38. Endo M, Sugiyama H (2018) DNA origami nanomachines. Molecules 23:1766

DNA-Based Nanoelectronics

5

"To make it smaller and smaller"—this is the leading trend of modern microelectronics that demands the shrinking of electrical circuits as an inevitable consequence of the increasing complexity of modern electronic devices. The handheld smartphone, combining a cell phone and a minicomputer, is a good example of that. Another example is lab-on-a-chip, a tiny multicomponent electronic microfluidic device of the size of a coin, which can be integrated with a smartphone to quickly detect multiple pathogens in a droplet of blood [1]. Evidently, if not to make electrical and electronic components smaller, the size of these and other new electronic devices could be enormous.

As a long-term trend, the tendency to shrink microelectronic devices more and more was predicted in 1965 by Gordon Moore, co-founder of the Intel Corporation, who described it in his famous paper "Cramming more components onto integrated circuits" [2]. In this paper, Moore noted that the number of components in electronic microcircuits had doubled every year from 1958 (the year of invention of electronic microcircuits, aka integrated circuits or microchips, by the Nobelist Jack S. Kilby) until 1965. And Moore reasonably anticipated that the trend for increasing complexity of electronic devices would continue for at least 10 years—hence their unavoidable miniaturization.

In fact, the validity of this trend, now known as "Moore's law," has been proved for half a century, and only in 2015 the pace of miniaturization of electronic devices had started to somewhat slow down due to space constraints imposed by physical limitations of silicon transistors, the major semiconductor components in modern electronics, and also bulky metallic lines used to connect various components of electronic microchips [3, 4]. However, experts think that the use of alternative materials that may replace the traditional silicon-based semiconductors and metallic wires in microchips could restore the exponential growth of integrated circuit complexity again [4, 5].

To this purpose, molecular electronics (aka nanoelectronics) has recently emerged, which involves the study and application of molecular building blocks for the fabrication of electronic components at the atomic scale, including the use of

© The Author(s), under exclusive licence to Springer Nature Switzerland AG 2020
V. V. Demidov, *DNA Beyond Genes*,
https://doi.org/10.1007/978-3-030-36434-2_5

conductive and semiconductive elements based on single polymeric molecules called molecular wires [6–8]. Such nano-sized conductive wires would serve as the electric energy-delivering "blood vessels" in nanoelectronic devices that could overcome the space/size limitations of conventional integrated circuits (Fig. 5.1). Scientists also contemplate the design of single-molecule transistors to be used as power amplifiers and electrical switches in nanoelectronic circuits [9, 10].

5.1   DNA as a Conductive Molecular Wire

In a search for conductive molecular wires among various promising organic polymers, scientists and engineers turned now to DNA molecules [11–14], the fundamental chemical basis of life, that can be obtained today by automated chemical synthesis with high yield and purity—essential prerequisite for their use as basic components of electronic devices. The reason for this is twofold. Firstly, in contrast to traditional manual or robotic assembly of common macroscale electronic devices, the feasibility of nanoelectronics will evidently depend heavily on the ability of molecular electronic components to direct and to form desired electronic circuits (or even a complete system) by themselves, since mechanical tools for precisely manipulating multiple single molecules might be costly and time/labor consuming. In this

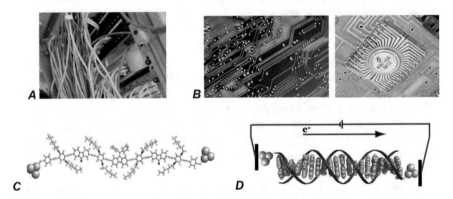

Fig. 5.1 Electro-wires: from macro to micro and nano. (**a**) Electro-wire is a major element of any electronic device, and reducing the size of a wire will evidently allow making the device more compact. (**b**) Integrated circuits are the smallest millimeter-size common electronic devices today with conductive pathways of copper sheets laminated onto a nonconductive substrate and with aluminum or gold bond wires, which are bonded to pads. (**c**) A single-molecule nanowire with a length of 7 nm. It is a rigid and almost linear oligomeric organic molecule made in 2006 by a team of British scientists with thiol groups at both ends, allowing the wire to dock between two gold electrodes [8]. (**d**) Schematics of imaginary molecular wire made of DNA and placed between two electrodes. The DNA double-helical structure facilitates charge transport: two sugar-phosphate DNA backbones (purple ribbons) are wrapped around the planar pairs of aromatic bases (blue), which tightly stack with each other, so that their outermost electron shells (called π orbitals) are partly overlapping one another. This overlap makes it possible for electrons to move between the neighboring bases, which could form a conductive core, a π-band, along the helical axis under certain conditions

regard, the unique capability of DNA molecules to be pre-programmed for self-directed assembly into composite constructs and complex networks, as it was discussed in the preceding chapter, would make these macromolecules a prime choice for future nanoelectronic technologies provided that the highly conductive DNA-based wires, robust DNA-based transistors, and other DNA-based components of electronic circuits could be designed [15].

Secondly, the steadily growing interest in the so-called DNA electronics is due to exceptional DNA double-helical structure, which from electrical viewpoint looks like a protective shell formed by the two intertwined threads wrapped around a conductive cord (Fig. 5.1d). Accordingly, some researchers believe that synthetic DNAs could be good conductors of electricity owing to delocalized π electrons of the stacked bases that may function as a quite efficient means for electrical conductivity, if conditions for long-range electron transfer along the DNA molecule were found [16].

Yet, although the issue of DNA conductivity has quite a long research prehistory,[1] conductance measurements done by different researchers on duplex DNA yielded a wide range of seemingly contradictory results (reviewed in [11–13]): some researchers claim that DNA samples they studied do act as semiconductors, or metallic-like conductors, or even as superconductors at very low temperatures [19], whereas the others report quite the opposite, i.e., DNA impedes the free flow of electrons, therefore acting as an electrical insulator.

These apparent contradictions have been attributed to the DNA sample differences, such as dissimilar DNA base sequences and lengths, and to diverse experimental conditions under which DNA samples were prepared and studied, including variations in temperature, water content, counterions, electrode contacts, and so on, any of which could affect DNA conducting properties [20]. Anyway, the growing consensus is that it is indeed possible to flow charges along relatively short, about 100 bp in length or less ($\leq$40 nm), DNA duplexes, although their conductivity is rather poor.[2] However, charge transport through longer duplex DNA molecules would be evidently much more limited due to some irregularities caused by dynamical disordering and thermal structural fluctuations, which are gradually accumulated in long DNA wires at normal temperatures [23].

Therefore, besides the double-stranded DNAs, researchers are also studying electrical conduction in alternative DNA wires based on the assembled three- or

---

[1] The DNA electronics originated in the 1960s when Belgian and British scientists first studied the electrical properties of DNA molecules in the dry state and concluded that DNA behaves like a semiconductor [17, 18]. In fact, these early studies were the first "out-of-genome" thinking of DNA.

[2] A recent study has found that conductivity of ~100-bp-long DNA duplexes was substantially improved by using fullerene groups (aka buckyballs, the molecules of carbon in the form of a hollow sphere) to anchor DNA to a surface: the measured DNA conductivity in this case was close to that of typical semiconductors, though it was still several orders of magnitude less than conductivity of pure metals [21]. It was also reported that the conductivity of 10–20-bp-long DNA duplexes increased almost ten times when 80% of water in conductive cell was replaced by ethanol, which caused the transition of DNA from the common so-called B-form to more compact A-form [22].

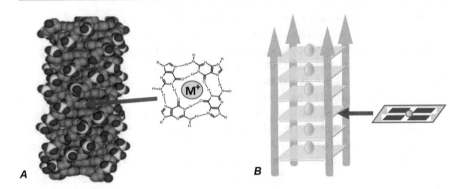

Fig. 5.2 Tetra-molecular G4 wire (aka DNA quadruplex). (**a**) Space-filling model of the G4-wire: it is the right-handed four-stranded helix assembled from the four parallel poly(dG) strands that are held together by hydrogen-bonded tetrads of guanine bases, one of which is shown at the right. The G4-tetrad embraces a monovalent metal ion (Li^+, Na^+, or K^+), which is coordinated by G's oxygen atoms and could provide extra stability to G4-tetrad. (**b**) Schematics of the G4-wire fragment with the stack of G4-tetrads shown as yellow parallelograms and an array of metal ions colored blue. One of the G4-tetrads is schematically portrayed at the right. Arrows indicate a uniform direction-ality of poly(dG) strands

four-stranded DNA helical structures [24–26], or in DNA mimics, such as non-charged peptide nucleic acid (PNA) [27]. The rationale for this is that the unconventional nucleic acid macromolecular structures may offer an improved stiffness and/or electronic overlap and delocalization that would enhance the conductivity of these macromolecules [28, 29].

The promising results were obtained by international team of researchers who studied the conductivity of so-called G4-DNA wires adsorbed on a mica substrate and composed of four strands of G nucleotides that run parallel to each other and form a braided 1D structure, with the guanine tetrad as a repeated unit (Fig. 5.2). The scientists found that when the applied electric potential across 300-bp-long (that is ~100 nm or 0.1 µm in length) G4-DNA wires was only a few volts, currents through these wires could reproducibly reach up to 100 pA [26]. They concluded that G4-DNA wires behave as semiconducting materials (their electrical conductance can be estimated as $\lesssim 10$ pS) and that the charge transport through them consists of multiple successive hopping between adjacent G tetrads, meaning the movement of an electron (or other charged particles) can be viewed as a series of "hops" or "jumps" from one G tetrad to another.

The observed robust long-distance charge transport through G4-wires can be attributed to a better π–π overlap, and therefore a higher degree of delocalization of the electronic states and consequently improved hopping, in G4-DNA than in duplex DNA [24, 28].[3] Moreover, the presence of metal ions at the core of the G4-DNA wires (Fig. 5.2) may additionally mediate charge motion, as it was observed with metallized duplex DNA.

---

[3] Note that superior conductivity of A-DNA duplexes was explained by longer π electron delocal-ization in A-DNA form compared to that in B-DNA [22].

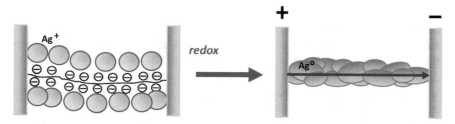

Fig. 5.3 Construction of DNA-templated silver wire connecting two electrodes. Left: Double-helical λ DNA bridging the two gold electrodes and loaded with silver ions via negatively charged phosphate groups of the DNA backbone. Right: The silver ion/DNA complex was reduced by redox reaction with hydroquinone, which resulted in the interelectrode DNA bridge covered by small metallic silver aggregates, thus forming a DNA-templated silver-coated wire. At the applied 1 V bias voltage, electric current through the silverized DNA nanowire (shown as a red arrow) reached 100 nA [30]

Indeed, DNA metallization was suggested 2 decades ago as another alternative to overcome poor electrical conductance of "naked" DNA molecules [30, 31] given the well-known ability of DNA duplexes to stably bind a variety of metal ions [32]. Using this ability, 100-nm-wide and more than 10-micron-long conductive silver wire was formed by loading dissolved silver ions onto the full-length DNA from bacteriophage λ, the virus infecting bacteria *E. coli* (Fig. 5.3) [30]. Remarkably, the electrical conductance of such a long DNA-templated wire was estimated as ~0.1 μS, which is more than 10,000 times larger than the conductance of much shorter G4-DNA wires described above, meaning a million times improvement in the DNA wire conductivity (see Glossary at the end of this book for explanation of terms). Still, the conductivity of silverized DNA nanowire is several orders of magnitude less than the conductivity of pure silver.

Besides fabrication of silver wires templated by DNA duplexes, chemical reduction of DNA-complexed metal ions was used for obtaining the DNA-templated nanowires of other metals, such as copper, platinum, and palladium [33–35]. Also, the old-style film photography technique for forming silver nanoclusters by illuminating the light-sensitive silver complexes (aka silver photography now largely replaced by digital photography) was exploited for DNA silverization using UV light for the in situ photoreduction of DNA-complexed silver ions [36]. Such a DNA-templated photoinduced silver deposition suits well the emerging nanoscale photolithography for fabricating nanoelectronic circuits [37].

Recently, the Israeli researchers have prepared highly conductive, almost micron-long gold-coated DNA nanowires by directly seeding DNA duplexes with gold nanoparticles, as an alternative to chemical reduction of ion/DNA complexes, and growing the DNA-attached nanoparticles further to form a continuous metallic path over DNA [38]. Thus produced metallized DNA wires were smoother than those previously reported, and their conductivity was stable for over a year and close to that of carbon nanowires—another promising material for future nanoelectronics [39].

Instead of using DNA duplexes as a core of metallized DNA wires, researchers from Duke University covered by silver the 25-nm-diameter DNA nanotubes self-assembled from the DNA triple-crossover tiles (see Fig. 3.7d in Chap. 3) [40]. Silverized DNA nanowires prepared this way demonstrated much higher electrical conductance compared to the silver-covered DNA duplexes [30], although their conductivity was slightly lower than that reported for palladium-covered DNA duplexes [35]. Also, the metallized DNA nanotubes are of much more uniform width and smooth appearance than the rather bumpy and grainy metallic nanowires templated on DNA duplexes, and they are generally overall thinner.

The use of linear DNA duplexes or 2D/3D DNA nanostructures as templates and scaffolds for the assembly of electronically active materials and chips is not limited to metallic species since conducting and semiconducting organic polymers can also be similarly employed. The very first design of that kind known to me was proposed in 1987 by Bruce Robinson and Nadrian Seeman, who suggested to employ nucleic acid-branched junctions (see Fig. 3.2a in Chap. 3) as scaffolds for docking conducting polymers [41]. The imaginary molecular scale memory device they hypothetically designed should allow an enormous density of information along with operation at electronic speeds over short distances. And though the estimated density of data storage in this device would be not as high as that in the DNA data storage devices described in Chap. 2, the access time should be much faster, i.e., on the order of picoseconds. Such a significant improvement in efficiency more than compensates for the extra size of the bit in the projected biochip.

While this bright idea still waits for its practical realization, the PPV semiconducting wires were already fabricated lying alongside the DNA molecules [15, 30]. PPV, or polyphenylene vinylene, is a diamagnetic polymer with low intrinsic electrical conductivity, which can be doped with iodine, ferric chloride, or alkali metals to form a variety of electrically conductive materials. Similar to fabrication of metallized DNA, the assembly of DNA-templated PPV wires was performed by self-attaching positively charged monomeric PPV precursors to the stretched negatively charged DNA strands and converting them to PPV polymers by a thermal treatment.

Recent study has proved that DNA, besides serving as a conductive nanowire itself or as a template for assembly of metallic nanowires, could also be employed for routing molecular wires in the user-defined paths. The international team of researchers from the USA, Germany, and Denmark have demonstrated that DNA molecules may work as sticky tapes helping to bend and to shape individual electrically conducting polymeric fibers into curve lines of any desired form by attaching them to specially designed 2D and 3D DNA-based scaffolds [42]. To do so, the DNA-grafted PPV polymers and DNA origami templates have been synthesized both featuring attached short single-stranded DNA oligonucleotides of defined nucleobase sequences that protrude across the polymer backbone and from the surface of DNA templates (Fig. 5.4; see Chap. 3 for DNA origami).

Since oligonucleotides attached to DNA scaffolds can be arranged in such a way that altogether they draw a predesigned route, the complementary oligonucleotides connected to PPV polymers will force them to arrange along the predesigned routings after DNA scaffolds and PPV polymers contact each other. The DNA-grafted PPV molecular wires, which can be more than 200 nm in length, and other electrically conducting polymers [43] might therefore be used to conductively connect electronic molecular circuitries of arbitrary geometries.

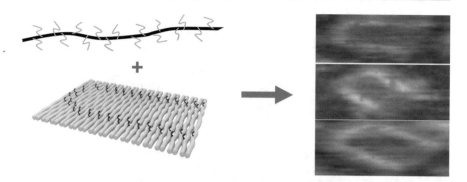

Fig. 5.4 Routing PPV polymers on DNA scaffolds. Left: Schematic illustration of DNA-grafted PPV polymer (top) and DNA origami scaffold (bottom) containing the U-shaped pattern of oligonucleotides (drawn in black) that are complementary to those attached to PPV (drawn in green). Right: Microscopic images of individual U-shaped PPV polymers after they were brought into contact with the corresponding DNA scaffolds. (Adapted from [42])

5.2 Towards DNA-Based Transistors and Diodes

One more possible field of DNA employment in nanoelectronics is the fabrication of molecular sized electronic components, e.g., transistors. Transistors are small semiconductor devices, which are essential parts of any electronic gadgets, and modern commercial transistors made of silicon may be about 50 nm in size or a bit smaller so that one electronic chip may have more than three billion transistors! Still, silicon transistors could not be shrunk significantly far more due to limits imposed on silicon by certain physical laws and phenomena, which would lead to large stochastic fluctuations disrupting proper functioning of a few nanometer-scale device [3, 4]. Therefore, microcircuit industry specialists are convinced that the next-generation, smaller size transistors will be built not from traditional silicon semiconductors, but from germanium, molybdenum, or carbon, which could be made in sizes of just a few nm [5, 44, 45]. And DNA may be an important player in the development of carbon-based transistors.

Carbon nanotubes (CNTs) could be one of the possible replacements for silicon that would shrink the size of transistors further. A CNT is a nanoscopic hollow cylinder with a one-atom wall width and with a nanometer-scale diameter, which is made up of carbon only. This is basically a rolled-up honeycomb mesh of carbon atoms, called graphene (see Fig. 5.5a). CNTs' very small dimensions and their metallic or semiconducting properties determined by both the graphene unique structure and the CNT's particular shape (mostly CNT's diameter and rolling orientation of graphene sheet) make them desirable building blocks for molecular electronics [46].

Indeed, in the pilot studies, CNTs have been used to create workable molecule-sized transistors, but it is rather challenging to accurately incorporate such tiny electronic devices into multicomponent integrated circuits [47]. The difficulty stems from the fact that CNTs lack recognition between molecular building blocks. Therefore, providing CNTs with this feature would greatly facilitate their correct

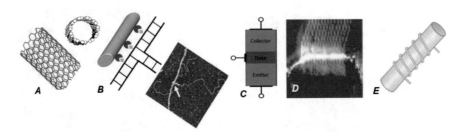

Fig. 5.5 Examples of using DNA in the design of CNT transistors. (**a**) Molecular structure of carbon nanotubes (side and front views), with carbon atoms being in wrapped hexagonal lattice. (**b**) Schematics of a DNA-templated nanotube (gray cylinder), which is attached to the three-way, T-shaped DNA junction via biotin-streptavidin links. Insert shows the micrograph of a CNT (thick line highlighted by light brown color) attached to a T-shaped DNA molecule—the three thin cross-ing lines highlighted by red color—with arrow showing the crossing/branching point. (Adapted from [49]; reproduced by permission of The Royal Society of Chemistry.) To form this complex, streptavidin protein (shown as red semispheres) is adsorbed on a CNT wall and DNA is modified with biotin (shown as small green squares). Streptavidin forms very strong bonds with biotin, therefore anchoring CNT to DNA. Arms of DNA junction can serve as wires to connect the entire structure to other electronic components and/or as sticky tapes to attach this structure to DNA scaf-fold. (**c**) Simplified diagram of a transistor showing that the overall geometry of DNA-templated CNT mimics the three terminals connecting transistor to an external circuit and making transistor to work as an amplifier: small current flowing from base to emitter causes large current to flow from collector to emitter. (**d**) The micrograph of a CNT (thick white line) grafted with DNA oligo-nucleotides and attached to a DNA origami scaffold (white mesh) containing complementary oli-gonucleotides, as it was assembled in [50]. (**e**) Schematics of a carbon nanotube wrapped with ssDNA by strong non-covalent interactions between them. DNA wrapping of CNTs can be used for fine-tuning of CNT's semiconducting properties

localization and interconnection within nanoscale self-assemblies bypassing the need for nanofabrication techniques and robotic nanomanipulations.

As in the case of DNA-grafted PPV molecular wires described above, coupling CNTs with DNA endows CNTs with molecular recognition capability, which enables the self-assembly of a functional nanoelectronic circuit comprising CNT transistors, their connection to other circuit components by DNA wires, and their targeted attachment to a scaffold DNA network [48–50]. Examples of the two such assemblies are shown in Fig. 5.5b–d: one is the construction of a DNA-templated CNT attached to the T-shaped, branched DNA molecule and another one is the con-struction of a DNA-grafted CNT attached to a DNA origami scaffold.

Furthermore, researchers found that under certain conditions a single-stranded (ss) DNA, which is more reactive than double-stranded DNA because of greater exposure of nitrogenous bases to the surrounding environment, would be readily adsorbed on a carbon nanotube, without the use of any other auxiliary tools, due to electrostatic and hydrophobic interactions between ssDNA and CNT [51, 52]. Consequently, being also much more flexible than double-stranded DNA, ssDNA strand becomes wound around carbon nanotube, as it is shown schematically in Fig. 5.5e. It was discovered afterward that the DNA-wrapped CNTs acquire modi-fied electronic transport characteristics which might be used for fine-tuning of CNT's semiconducting properties [52–54].

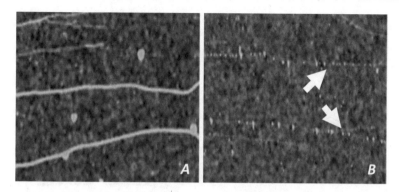

Fig. 5.6 Fabrication of graphene nanoribbons (GNRs) with the use of DNA templates. (**a**) Micrograph of a pair of DNA molecules impregnated with copper and stretched on a silicon wafer. (**b**) Micrograph of a pair of GNRs (pointed by white arrows) obtained by catalytical transformation of DNA heated and bathed in methane gas. (Adapted from [55])

Scientists from Stanford University suggested a different way to build a carbon-based molecular sized transistor with the help of DNA [55]. Instead of nanotubes, they decided to use nanoscaled thin, narrow, and long planar strips of single-carbon layers, called graphene nanoribbons (GNRs), which are also known to have promising semiconducting properties for high-performance molecular electronics [46]. The major challenge for GNRs to be used as one of the next-generation nanoelectronic materials is the current scarcity of effective technologies for industrial scale production of GNRs with the requisite nano-precision [56], in contrast to well-established procedures for growth of atomically precise carbon nanotubes [57, 58].

To cope with this challenge, the Stanford team came up with an idea of using long and stretched copper-metallized DNA molecules as chemical templates to grow on them nanoscopic ribbons of pure carbon, which would take on the shape and dimensions of a DNA template that is catalytically transformed into a 10-μm-long and 10-nm-wide GNR. To realize this idea, researchers first coated nanosized silicon wafer with the flow-stretched ~10-μm-long DNA molecules derived from bacteria. Then they immersed this chip into solution of copper salt to impregnate DNA with copper, which is known as a catalyst for growth of single-layer graphene (Fig. 5.6a). Next, the chip with copper-metallized DNA was dried and heat-treated in the presence of methane—natural gas—which contains carbon and hydrogen atoms. As a result, the DNA strands were chemically converted to graphitic nanoribbons with sizes of DNA templates (Fig. 5.6b).

The precise chemistry of this process is largely unknown, but researchers think that the heat and the catalyst released some of the carbon atoms in the DNA and methane, which stayed close to disintegrated DNA strands and quickly joined together to form nanoribbons that follow the form of these strands. Finally, electrical characterization of thus obtained GNRs proved that depending on the growth conditions, metallic or semiconducting graphitic nanoribbons could be formed and used to make working nanotransistors [55].

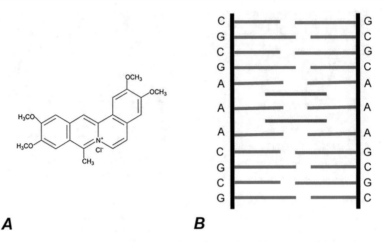

A   **B**

Fig. 5.7 Construction of a DNA diode. (**a**) Coralyne is a small planar crescent-shaped molecule known to intercalate duplex DNA. (**b**) Schematics of the intercalation complex formed by the two coralyne molecules (red lines) with undecameric DNA duplex containing three noncanonical A–A base pairs at the center (preferable sites for coralyne intercalation). This complex behaves as a molecular diode when applying voltage to its ends [61]

Other studies revealed that DNA itself acquires properties that could mimic the electronic behavior of conventional semiconductors, when complexed with special small molecules, termed intercalators, or when carried certain DNA modifications [59–61]. This suggests alternative ways for using DNA in building molecular transistors [59, 60] and other key electronic components, with DNA molecular diode (aka rectifier)—device that allows electric current to flow primarily in one direction—being one of them [61].

Unaided, DNA does not act like a diode; that is, it does not convert an alternating current to a direct one. However, when the Israeli-American team of researchers inserted a pair of DNA-intercalating molecules into certain points within DNA (see Fig. 5.7) and applied a voltage to this complex, electric current flowed through it more than ten times as strongly in one direction than the other [61]. Theoretical models revealed that this effect resulted from the intercalator-induced local spatial asymmetry in the distribution of electron states along the DNA chain, which, in turn, leads to directional preference in the charge transport through DNA molecules.

Though the ratio of the forward and backward currents in DNA diode, which is its primary performance parameter, is not as high as in other single-molecule diodes created so far [62, 63], this work offered a new strategy for engineering DNA-based electronic elements by exploiting DNA-small molecule interactions. Moreover, the scientists hope that by extending this study they can further improve their DNA diode to achieve functional molecular devices.

I believe that the contents of this chapter prove that there is a wide range of possibilities for prospective using DNA molecules in the imminent era of nanoelectronics—from forming conducting nanowires to fabrication of nanosized electronic components either made of DNA or templated and functionalized by DNA. As to

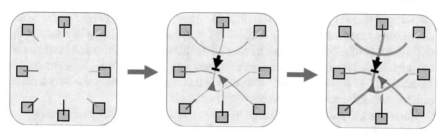

Fig. 5.8 Imaginary assembly of nanometer-size electronic circuit from the DNA-based building blocks. Left: Nanoscale inert substrate (aka wafer; shown enlarged in this schematics) carrying an array of square nanoelectrodes with attached short single-stranded DNAs (aka oligonucleotides), each having different terminal sequence (represented by different colors). Center: The DNA network is self-assembled by binding of DNA molecules to oligonucleotides and to DNA-based diode (shown as →I) and DNA-CNT transistors (shown as blue triangles). Right: "Metallization" of DNA network by chemical deposition of metal atoms onto DNA molecules with formation of electro-conductive DNA wires (shown as thick brown lines)

the last option, the capacity of DNA self-assembly for constructing nanoscale electronic devices is very attracting, given the fundamental limitations of conventional photolithography-based assembly of silicon circuits. Indeed, common photolithography is governed by the laws of classical optics so it is basically unable to create features significantly smaller than the wavelengths of light, which are in the range of several 100 nm. The emerging quantum lithography can go well below this limit, but it has so far unresolved problems with efficiency [64]. And even if a nanometer-sized silicon chip could be constructed by quantum lithography, it will be nonfunctional due to large stochastic fluctuations partly caused by random positions of doping atoms within the silicon substrate, as it was already mentioned above.

These are the reasons for many researchers to turn their attention to alternative electronic devices created from molecular structures, such as carbon nanotubes, or CNTs, in which the locations of all atoms are well defined. However, an outstanding problem is how to arrange CNTs into a desired functional pattern on a chip. And apparently only DNA self-assembly could effectively solve this problem by functionalizing CNTs with molecular recognition abilities.

Given all described above, the future DNA-based nanoelectronic assemblies could be viewed as it is shown in Fig. 5.8 (like it was suggested in [15]). First, an array of nanoelectrodes is manually or robotically fabricated on a nano-sized wafer surface by common techniques using standard microfabrication tools [39, 65, 66] and each electrode in an array is functionalized with specific "anchor" oligonucleotides using well-known chemistry [8, 30]. Next, the wafer surface is covered by a nanodrop of solution containing a variety of long distinctive double-stranded DNA molecules with overhanging sticky ends,[4] as well as various DNA-based nanoelectronics components, like DNA diodes and DNA-CNT transistors. Some sticky DNA

---

[4]Long DNA duplexes with sticky ends are commonly created by restriction endonucleases (aka restriction enzymes), some of which can cut the two DNA strands at specific sites four-five base pairs from each other, creating a 5′ overhang in one molecule and a complementary 5′ overhang in the other.

ends should match specific electrode-bound oligonucleotides, whereas others would match the protruding termini of DNA-CNT transistors, etc. This will result in the self-assembly of a DNA network having well-defined connectivity and holding certain electronic elements at predetermined positions. Finally, the DNA network is "metallized," as it was described above, to become a net of conductive wires, therefore forming a functional nanometer-size electronic circuit.

Currently, DNA electronics keeps a high promise, but only future can show us if it will come to practical life in the modernized integrated circuits industry. Surprisingly, though there is no one presently known biological function that requires DNA to be an effective conduit for long-range charge transfer, some scientists seriously consider that Nature may already use DNA conductivity in certain systems which have yet to be discovered [67, 68]. And I hope that forthcoming studies will soon reveal to us if this is true or not.

References

1. Chen W et al (2017) Mobile platform for multiplexed detection and differentiation of disease-specific nucleic acid sequences, using microfluidic loop-mediated isothermal amplification and smartphone detection. Anal Chem 89:11219–11226
2. Moore GE (1965) Cramming more components onto integrated circuits. Electronics 38:114–117
3. Wang Y, Han R, Liu X, Kang J (1998) The challenges for physical limitations in Si microelectronics. In: Solid-state and integrated circuit technology, pp 25–30, IEEE Press; Beijing, China
4. Keyes RW (2005) Physical limits of silicon transistors and circuits. Rep Prog Phys 68:2701–2746
5. Peide DY (2016) Switching channels. IEEE Spectr 53(12):40–45
6. Robertson N, McGowan CA (2003) A comparison of potential molecular wires as components for molecular electronics. Chem Soc Rev 32:96–103
7. Guldi DM, Nishihara H, Venkataraman L (2015) Molecular wires. Chem Soc Rev 44:842–844
8. Ashwell GJ et al (2006) Single-molecule electrical studies on a 7 nm long molecular wire. Chem Commun 45:4706–4708
9. Perrin ML, Burzuri E, van der Zant HSJ (2015) Single-molecule transistors. Chem Soc Rev 44:902–919
10. Park H et al (2000) Nanomechanical oscillations in a single-C60 transistor. Nature 407:57–60
11. Taniguchi M, Kawai T (2006) DNA electronics. Phys E 33:1–12
12. Triberis GP, Dimakogianni M (2009) DNA in the material world: electrical properties and nano-applications. Recent Pat Nanotechnol 3:135–153
13. Abdalla S (2011) Electrical conduction through DNA molecule. Prog Biophys Mol Biol 106:485–497
14. Wang K (2018) DNA-based single-molecule electronics: from concept to function. J Funct Biomater 9:e8
15. Eichen Y, Braun E, Sivan U, Ben-Yoseph G (1998) Self-assembly of nanoelectronic components and circuits using biological templates. Acta Polym 49:663–670
16. Bixon M et al (1999) Long-range charge hopping in DNA. Proc Natl Acad Sci U S A 96:11713–11716
17. Duchesne J et al (1960) Thermal and electrical properties of nucleic acids and proteins. Nature 188:405–406
18. Eley DD, Spivey DI (1962) Semiconductivity of organic substances. 9. Nucleic acid in dry state. Trans Faraday Soc 58:411–417
19. Kasumov AY et al (2001) Proximity-induced superconductivity in DNA. Science 291:280–282

20. Porath D, Cuniberti G, Di Felice R (2004) Charge transport in DNA-based devices. Top Curr Chem 237:183–227
21. Jimenez-Monroy KL et al (2017) High electronic conductance through double-helix DNA molecules with fullerene anchoring groups. J Phys Chem A 121:1182–1188
22. Artes JM et al (2015) Conformational gating of DNA conductance. Nat Commun 6:8870
23. Yu ZG, Song X (2001) Variable range hopping and electrical conductivity along the DNA double helix. Phys Rev Lett 86:6018–6021
24. Haruna K et al (2004) The effect of triple helix forming on charge transport in DNA. Nucleic Acids Symp Ser 48:241–242
25. Woiczikowski PB et al (2010) Structural stability versus conformational sampling in bio-molecular systems: Why is the charge transfer efficiency in G4-DNA better than in double-stranded DNA? J Chem Phys 133:035103
26. Livshits GI et al (2014) Long-range charge transport in single G-quadruplex DNA molecules. Nat Nanotechnol 9:1040–1046
27. Venkatramani R et al (2011) Evidence for a near-resonant charge transfer mechanism for double-stranded peptide nucleic acid. J Am Chem Soc 133:62–72
28. Esguerra M, Nilsson L, Villa A (2014) Triple helical DNA in a duplex context and base pair opening. Nucleic Acids Res 42:11329–11338
29. Felice D et al (2005) Strain-dependence of the electronic properties in periodic quadruple heli-cal G4-wires. J Phys Chem B 109:22301–22307
30. Braun E, Eichen Y, Sivan U, Ben-Yoseph G (1998) DNA-templated assembly and electrode attachment of a conducting silver wire. Nature 391(6669):775–778
31. Aich P et al (1999) M-DNA: a complex between divalent metal ions and DNA which behaves as a molecular wire. J Mol Biol 294:477–485
32. Spiro TG (ed) (1980) Nucleic acid–metal ion interactions. Wiley Interscience, New York
33. Monson CF, Woolley AT (2003) DNA-templated construction of copper nanowires. Nano Lett 3:359–363
34. Seidel R et al (2004) Synthesis of platinum cluster chains on DNA templates: conditions for a template-controlled cluster growth. J Phys Chem B 108:10801–10811
35. Richter J, Mertig M, Pompe W (2001) Construction of highly conductive nanowires on a DNA template. Appl Phys Lett 78:536–538
36. Berti L, Alessandrini A, Facci P (2005) DNA-templated photoinduced silver deposition. J Am Chem Soc 127:11216–11217
37. Seisyan RP (2011) Nanolithography in microelectronics: a review. Tech Phys 56:1061–1073
38. Stern A et al (2018) Highly conductive thin uniform gold-coated DNA nanowires. Adv Mater 30:e1800433
39. Lim Y et al (2017) Increase in graphitization and electrical conductivity of glassy carbon nanowires by rapid thermal annealing. J Alloys Compounds 702:465–471
40. Liu D, Park SH, Reif JH, LaBean TH (2004) DNA nanotubes self-assembled from triple-crossover tiles as templates for conductive nanowires. Proc Natl Acad Sci U S A 101:717–722
41. Robinson BH, Seeman NC (1987) The design of a biochip: a self-assembling molecular-scale memory device. Protein Eng 1:295–300
42. Knudsen JB et al (2015) Routing of individual polymers in designed patterns. Nat Nanotechnol 10(10):892–898
43. Walton DJ (1990) Electrically conducting polymers. Mater Des 11:142-152
44. Avouris P, Chen Z, Perebeinos V (2007) Carbon-based electronics. Nat Nanotechnol 2:605–615
45. Desai SB et al (2016) MoS$_2$ transistors with 1-nanometer gate lengths. Science 354:99–102
46. Soldano C, Talapatra S, Kar S (2013) Carbon nanotubes and graphene nanoribbons: potentials for nanoscale electrical interconnects. Electronics 2:280–314
47. Javey A et al (2002) Carbon nanotube transistor arrays for multistage complementary logic and ring oscillators. Nano Lett 2(9):929–932
48. Keren K et al (2003) DNA-templated carbon nanotube field-effect transistor. Science 302(5649):1380–1382

49. Lyonnais S et al (2009) A three-branched DNA template for carbon nanotube self-assembly into nanodevice configuration. Chem Commun 6:683–685
50. Maune HT et al (2010) Self-assembly of carbon nanotubes into two-dimensional geometries using DNA origami templates. Nat Nanotechnol 5(1):61–66
51. Zheng M et al (2003) DNA-assisted dispersion and separation of carbon nanotubes. Nat Mater 2:338–342
52. Hwang JS et al (2008) Electronic transport characteristics of a single wall carbon nanotube field effect transistor wrapped with deoxyribonucleic acid molecules. J Phys Conf Ser 109:012015
53. Cha M et al (2009) Reversible metal-semiconductor transition of ssDNA-decorated single-walled carbon nanotubes. Nano Lett 9:1345–1349
54. Lu Y et al (2006) DNA functionalization of carbon nanotubes for ultrathin atomic layer deposition of high κ dielectrics for nanotube transistors with 60 mV/decade switching. J Am Chem Soc 128(11):3518–3519
55. Sokolov AN et al (2013) Direct growth of aligned graphitic nanoribbons from a DNA template by chemical vapour deposition. Nat Commun 4:2402
56. Chen Q, Ma L, Wang J (2016) Making graphene nanoribbons: a theoretical exploration. WIREs Comput Mol Sci 6:243–254
57. Eatemadi A et al (2014) Carbon nanotubes: properties, synthesis, purification, and medical applications. Nanoscale Res Lett 9(1):393
58. Purohit R et al (2014) Carbon nanotubes and their growth methods. Procedia Mater Sci 6:716–728
59. Maruccio G et al (2003) Field effect transistor based on a modified DNA base. Nano Lett 3:479–483
60. Behnia S, Fathizadeh S, Ziaei J (2017) Controlling charge current through a DNA based molecular transistor. Phys Lett A 381:36–43
61. Guo C et al (2016) Molecular rectifier composed of DNA with high rectification ratio enabled by intercalation. Nat Chem 8:484–490
62. Capozzi B et al (2015) Single-molecule diodes with high rectification ratios through environmental control. Nat Nanotechnol 10:522–527
63. Chen X et al (2017) Molecular diodes with rectification ratios exceeding 105 driven by electrostatic interactions. Nat Nanotechnol 12:797–803
64. Kothe C, Bjork G, Inoue S, Bourennane M (2011) On the efficiency of quantum lithography. New J Phys 13:043028
65. Menon VP, Martin CR (1995) Fabrication and evaluation of nanoelectrode ensembles. Anal Chem 67:1920–1928
66. Brevnov DA et al (2004) Patterning of nanoporous anodic aluminum oxide arrays by using sol−gel processing, photolithography, and plasma etching. Chem Mater 16:682–687
67. Merino EJ, Boal AK, Barton JK (2008) Biological contexts for DNA charge transport chemistry. Curr Opin Chem Biol 12:229–237
68. Grodick MA, Muren NB, Barton JK (2015) DNA charge transport within the cell. Biochemistry 54:962–973

Concluding Remarks

6

When James Watson and Francis Crick described in a succinct 1953 report to *Nature* magazine their model of the DNA structure—the double helix—they clearly noted its importance for storing genetic information in a witty concluding remark [1]: "It has not escaped our notice that the specific pairing we have postulated immediately suggests a possible copying mechanism for the genetic material." Namely this specific pairing of DNA nucleobases with the unique double-helical shape of DNA molecules, which looks much like a spirally twisted ladder with two rails counterparting each other and being capable of coming apart like a zipper to enzymatically replicate this ladder in two identical copies, gives DNA the power to serve with great precision as a universal genetic information carrier for all living beings on Earth.

Yet, the contents of this book prove that DNA special structure and related exceptional physicochemical DNA properties also make these molecules extremely attracting for lots of other uses, too. As a result of this attraction, we are witnessing these days the emergence of fields of structural DNA nanotechnology and DNA computing, DNA electronics and DNA data storage, and rise of DNA machines, all presented above in my book with anticipation that its readers will ultimately become more familiar with "uncommon" DNA.

But this is not all: many new fields of DNA employment beyond the genome are just budding, such as the DNA-guided and DNA-catalyzed chemical syntheses, DNA-templated magnetic and optical devices, and DNA-based environmental detoxicants [2–7], which are waiting a new book for them to be covered after their full development from their currently embryonic state. Besides DNA, several pilot studies show that the high-affinity DNA mimics, such as peptide nucleic acids (PNAs), can also be employed for various auxiliary functions to further extend the potential of DNA in a material world [8–12].

Moreover, inspired by the success of structural DNA nanotechnology, the DNA's sister molecule RNA (*ribo*nucleic *a*cid) is coming now to the front line of biotech research as a key player in the emerging field of RNA nanotechnology [13]. One of the reasons behind this rise is a rich inherent architectural potential of RNA, which unlike DNA can use a variety of non-Watson-Crick base pairs to form diverse

© The Author(s), under exclusive licence to Springer Nature Switzerland AG 2020
V. V. Demidov, *DNA Beyond Genes*,
https://doi.org/10.1007/978-3-030-36434-2_6

nanostructures that cannot be built from DNA. Another important advantage of RNA over DNA is that any desired RNA might be encoded in synthetic DNA and produced in multiple copies inside of cells from corresponding recombinant genes.

This means that the self-folded RNA nanostructures could be formed intracellularly by DNA-to-RNA transcription, as it was indeed demonstrated by Caltech researchers, who cloned in yeast cells synthetic RNA nanostructures functioning as logic gates and switching intracellular molecular inputs into increased or decreased gene expression outputs, thus controlling certain cellular functions [14]. And recent study has shown that there are no limits to the variety and complexity of DNA-encoded RNA nanostructures produced and self-folded inside live cells—the joint team of American and Chinese researchers were able to grow and to harvest from bacteria RNA nanostructures with the shapes of five-petal flower, tetrahedron, and tetra-square [15]. Meanwhile, the joint team of ASU and NYU researchers proved that artificial DNA nanostructures can also be cloned and amplified in microbial cells by using genetically engineered viruses [16]. These pilot studies with in vivo-replicable DNA and RNA nanostructures provide strong evidence that it is principally possible to move the nucleic acid nanotechnology into living cells and organisms, and they therefore somewhat justify the concern first expressed by Eric Drexler, the founder of the Foresight Institute,[1] in his landmark 1986 book [17], where he cautioned about the potential threat posed by nanoscale replicating entities—essentially artificial life forms.[2]

Indeed, the in vivo-encoded artificial self-assembling DNA and/or RNA devices (let us call them DNA/RNA nanoreplicators), which have the capacity to replicate and also are capable of manipulating genes within live cells, as it was already verified by different teams of researchers [14–16, 18, 19], can randomly mutate and randomly recombine with cellular genetic elements in the same natural way as it occurs with other intracellular nucleic acids. As such, DNA/RNA nanoreplicators could evolve to acquire new hazardous abilities, and they may also escape, like infectious viruses always do, into the outer environment quickly getting out of control. Just imagine what might happen next ….

But do not hold your breath! And fortunately there is no need to keep your fingers crossed to avoid the worst: all these recombinant DNA/RNA nano-constructs are not autonomous ones, as they require live cells for their reproduction. So they can be well controlled by following the National Institutes of Health (NIH) Guidelines for

---

[1] Foresight Institute is a leading nonprofit organization researching and fostering technologies of fundamental importance for the human future, focusing on molecular machine nanotechnology, cybersecurity, and artificial intelligence.

[2] Drexler's original worry, shared later by other scientists, was about the feasible danger of replicating abiotic or mechanical biovorous nano-assemblers—parts of imaginary in vitro nanoscale factories capable of reproducing themselves—which could be developed in the lab by future nanotechnologists, and could then accidentally get away into the living world, where they consume the entire biosphere in a ravenous quest for fueling food, after which nothing remained there but an immense artificial mass of nano-assemblers he called "gray goo." Yet, this hypothetical gray goo scenario is well applicable to artificial biotic nanoreplicators, too.

Research Involving Recombinant DNA Molecules[3] and similar regulations in other countries engaged in such experiments, which are mandatory for any research activity involving recombinant and synthetic nucleic acid molecules [20]. Therefore, we should feel secure with all these innovations since almost 50 years of worldwide experience with recombinant DNAs has proved complete safety of this technology when researchers follow these rules.[4]

And it is hard to envision how artificial DNA/RNA nanoreplicators could become fully autonomous to reproduce themselves from scratch, that is, from the molecular milieu outside cells without any other aids. Even though DNA and RNA molecular constructs have recently been designed that self-replicate in vitro, they can reproduce themselves only from a number of sufficiently large and complex pre-prepared parts by connecting them together [21, 22]. Hence, there remains an unfilled gap in the autonomy of these nanoreplicators when compared with living organisms.

Still, one may imagine that the runaway nonautonomous DNA/RNA nanoreplicators could somehow further evolve into autonomously replicating entities by exploiting the DNA/RNA self-replicating and catalytic capabilities [3, 21], along with unlimited structural and informational potentials of DNA, its robust robotic capabilities, and microelectronic abilities as described in this book. This might finally give the origin of alternative nanobiobot-like forms of life, which could be highly competitive with presently living organisms, thus threatening their existence. Yet, based on our current knowledge of how life on Earth started and evolved [23, 24], even the rise of some primitive autonomous DNA/RNA replicators—ancestors of kind of life forms—most probably would require millions and millions of years to pass.

Therefore, the self-development of alternative DNA/RNA-based autonomously replicating entities—whichever they are—needs a very long time to proceed, and it is highly unlikely that such an event will remain absolutely unnoticed by human beings to take certain preventive and protective measures, if necessary. For all these reasons, I believe that the imaginable danger from evolving artificial DNA/RNA escapees lies well in the realm of science fiction.

And returning to the major topic of this book—DNA as a multipurpose engineering material—I would like to think that, as in any growing field, the most exciting developments in the DNA's material world are yet to come. So I am finishing this book with a nice concluding phrase from the notable article co-authored by Francis Crick [25]: "DNA is such an important molecule that it is almost impossible to learn too much about it."

---

[3] The NIH Recombinant DNA Guidelines were first issued in 1976 in the *Federal Register*, and they are overseen and continually revised by the NIH Recombinant DNA Advisory Committee, who reviewed all related research activities in accordance with these guidelines.

[4] There are many additional dimensions of safety controls that can be engineered into designs for replicators, as discussed in the Foresight Guidelines for Responsible Nanotechnology Development (https://foresight.org/guidelines/current.php#Replicators).

References

1. Watson JD, Crick FHC (1953) Molecular structure of nucleic acids: a structure for deoxyribose nucleic acid. Nature 171:737–738
2. Linko V et al (2016) DNA-based enzyme reactors and systems. Nanomaterials 6:E139
3. Silverman SK (2016) Catalytic DNA: scope, applications, and biochemistry of deoxyribozymes. Trends Biochem Sci 41:595–609
4. Watson SM, Mohamed HD, Horrocks BR, Houlton A (2013) Electrically conductive magnetic nanowires using an electrochemical DNA-templating route. Nanoscale 5:5349–5359
5. Dugasani SR et al (2013) Magnetic characteristics of copper ion-modified DNA thin films. Sci Rep 3:1819
6. Kawabe Y, Sakai KI (2011) DNA based solid-state dye lasers. Nonlin Optics Quant Optics 42:273–282
7. Hu X, Mu L, Zhou Q, Wen J, Pawliszyn J (2011) ssDNA aptamer-based column for simultaneous removal of nanogram per liter level of illicit and analgesic pharmaceuticals in drinking water. Environ Sci Technol 45:4890–4895
8. Nulf CJ, Corey DR (2002) DNA assembly using bis-peptide nucleic acids (bisPNAs). Nucleic Acids Res 30:2782–2789
9. Kuhn H, Cherny DI, Demidov VV, Frank-Kamenetskii MD (2004) Inducing and modulating anisotropic DNA bends by pseudocomplementary peptide nucleic acids. Proc Natl Acad Sci U S A 101:7548–7553
10. Breitenstein M, Nielsen PE, Holzel R, Bier FF (2011) DNA-nanostructure-assembly by sequential spotting. J Nanobiotechnol 9:54
11. Flory JD et al (2014) Purification and assembly of thermostable Cy5 labeled γ-PNAs into a 3D DNA nanocage. Artif DNA PNA XNA 5:1–8
12. Pedersen RO, Kong J, Achim C, LaBean TH (2015) Comparative incorporation of PNA into DNA nanostructures. Molecules 20:17645–17658
13. Jasinski D, Haque F, Binzel DW, Guo P (2017) Advancement of the emerging field of RNA nanotechnology. ACS Nano 11:1142–1164
14. Win MN, Smolke CD (2008) Higher-order cellular information processing with synthetic RNA devices. Science 322:456–460
15. Li M et al (2018) In vivo production of RNA nanostructures via programmed folding of single-stranded RNAs. Nat Commun 9:2196
16. Lin C et al (2008) In vivo cloning of artificial DNA nanostructures. Proc Natl Acad Sci U S A 105:17626–17631
17. Drexler KE (1986) Engines of creation: the coming era of nanotechnology. Anchor Press/Doubleday, New York
18. Vogel M, Weigand JE, Kluge B, Grez M, Suess B (2018) A small, portable RNA device for the control of exon skipping in mammalian cells. Nucleic Acids Res 46:e48
19. Groher F et al (2018) Riboswitching with ciprofloxacin—development and characterization of a novel RNA regulator. Nucleic Acids Res 46:2121–2132
20. US Congress Office of Technology Assessment Report (1984). Commercial Biotechnology: An International Analysis. Appendix F: Recombinant DNA Research Guidelines, Environmental Laws, and Regulation of Worker Health and Safety, US Government Printing Office
21. Robertson MP, Joyce GF (2014) Highly efficient self-replicating RNA enzymes. Chem Biol 21:238–245
22. Kim J, Lee J, Hamada S, Murata S, Park SH (2015) Self-replication of DNA rings. Nat Nanotechnol 10:528–533
23. Dodd MS et al (2017) Evidence for early life in Earth's oldest hydrothermal vent precipitates. Nature 543:60–64
24. Betts HC et al (2018) Integrated genomic and fossil evidence illuminates life's early evolution and eukaryote origin. Nat. Ecol Evol 2:1556–1562
25. Crick FHC, Wang JC, Bauer WR (1979) Is DNA really a double helix? J Mol Biol 129:449–461

Glossary

3′-End and 5′-end   A single-stranded noncircular DNA molecule has two non-identical ends, the 3′-end and the 5′-end (usually pronounced "three prime end" and "five prime end"). These numbers refer to the numbering of carbon atoms in the deoxyribose, which is a sugar part of the DNA backbone. In the backbone of DNA the 5′ carbon of one deoxyribose is linked to the 3′ carbon of another by a phosphate group. The 5′ carbon of this deoxyribose is again linked to the 3′ carbon of the next, and so forth, thus imposing a directionality (or polarity) on DNA strands. Accordingly, the arrow on the DNA strand schematics usually goes in 5′ to 3′ direction. Importantly, DNA polymerase can only add new nucleotides to a 3′-end of a growing chain.

A-DNA   This is one of the possible double-stranded structures which DNA can adopt. It is a right-handed double helix somewhat similar to the more common B-DNA form, but with a shorter, more compact helical structure whose base pairs are not perpendicular to the helix axis as in B-DNA. DNA is driven into the A form under dehydrating conditions, when DNA is in crystals or in dry films. This was the first discovered distinct DNA conformation and that is why it was named so.

Algorithm   It is a sequence of explicit instructions or a set of specific rules that are followed step by step to solve a problem and/or complete a task.

Antiparallel strands   DNA strands are antiparallel if they run alongside each other but in opposite directions (see the entry "3′-end and 5′-end" above).

B-DNA   The prevailing right-handed double-helical form of DNA in aqueous solutions under physiological conditions. Actually, B-DNA is not a well-defined conformation but a family of related DNA conformations that occur at the high hydration levels normally present in living cells. In B-DNA, the base pairs are perpendicular to the helix axis and it is also characterized by a smooth winding of sugar-phosphate backbone. When dehydrated, B-DNA transforms to A-DNA.

Base pair   A pair of complementary bases in a double-stranded nucleic acid molecule, consisting of a purine in one strand linked by hydrogen bonds to a pyrimidine in the other. In double-stranded DNA, cytosine (C, pyrimidine) always pairs with guanine (G, purine), and thymine (T, pyrimidine) with adenine (A, purine).

Binary digit or bit   This is one of the two digits (0 or 1) used in a binary system of mathematical notation. The term *bit* is a linguistic blend of the words *bi*nary

© The Author(s), under exclusive licence to Springer Nature Switzerland AG 2020
V. V. Demidov, *DNA Beyond Genes*,
https://doi.org/10.1007/978-3-030-36434-2

and digi*t*. A bit can have only one of the two values, and may therefore be physically implemented with a two-state device. It is the smallest unit of information in a computer used for storing digital data and it has a logical value equivalent of true/false, yes/no, or on/off.

Binary number   In mathematics and digital electronics, a binary number is a number expressed in the binary numeral system, which uses only two digital symbols: typically "0" (zero) and "1" (one). Each digit is referred to as a bit. For example, counting from 0 to 10 in binary looks like this: 0, 1, 10, 11, 100, 101, 110, 111, 1000, 1001, 1010. Because of its straightforward implementation in digital electronic circuitry using logic gates, the binary system is used by almost all modern computers and computer-based devices.

Biotin (aka vitamin B7 or vitamin H)   This is a small molecule that binds tightly to tetrameric proteins avidin and streptavidin (up to four biotin residues per single protein). This is often used in different biotechnological applications, some of which are presented in this book.

Circular DNA   It is the DNA that forms a closed loop and has no ends; circular DNA can be either in a single-stranded form or in a double-stranded form with the double helix looping around to make a complete circle.

Complementarity/complementary   In terms of DNA, complementarity is the ability of the two nucleic acid bases (aka nucleobases or nitrogenous bases) to form matching Watson-Crick base pairs; that is, adenine (A) is complementary to thymine (T) and guanine (G) is complementary to cytosine (C), and vice versa. Accordingly, two DNA sequences are complementary to each other if their opposite alignment shows that they are composed of matching base pairs. For instance, the complementary sequence to G-T-A-C is C-A-T-G.

Conductance and conductivity   Both represent the ability of an object or the property of an object's material to conduct electric current, accordingly. They are measured correspondingly in siemens (symbol: S) and siemens per meter (S/m, as specific conductance). Their high values indicate an object or a material that readily allows the flow of electric current (called conductors or metals). Conversely, their low values indicate an object or a material that resists the flow of electric current (called isolators).

Covalent bonds   Also called molecular bonds, these are the strong chemical bonds that form molecules by sharing electron pairs between atoms.

Digital microfluidics (DMF)   This is a technique for manipulation of discrete droplets on a substrate, which allows timed motion of fluids with accurate control, representing an alternative to the conventional microfluidics where fluids move in enclosed channels.

Diode (aka rectifier)   A two-terminal electronic component that conducts electric current primarily in one direction. Diodes are mostly made of silicon, but other semiconducting materials such as selenium and germanium can also be used. They perform many different functions in common microelectronic devices.

Directionality (aka polarity) of DNA   See the entry "3′-end and 5′-end" above.

DNA hybridization/annealing See below the entry "Hybridization."

DNA ligases These are the enzymes that covalently join the two oligo- or poly-nucleotide DNA strands together when their ends are in close proximity, thus creating a larger continuous DNA strand.

DNA polymerases These are the enzymes that perform polymerization of DNA units, aka nucleotides, by sequentially adding them into a growing DNA polynucleotide strand. Most of the DNA polymerases employ another DNA strand as a template to add into a new DNA polynucleotide strand only those nucleotides that match the template DNA sequence. Thus, the major "job" of DNA polymerases is the DNA duplication, in which the polymerase "reads" a sequence of nucleotides comprising the template DNA single strand to use it for synthesis of a novel DNA strand. This process results in another piece of DNA, which is complementary to the template DNA strand and identical to the template's previous partner strand in the original DNA duplex.

DNA sequence This is an order of the four possible nucleotide bases, abbreviated as A, T, G, or C, in a particular DNA molecule. For example, it could be G-T-A-C-G-A-A-C-T or T-C-T-A-C-G-G-T-T for a 9-mer sequence read. Since strands have a directionality, DNA sequences are usually written in the 5′ -> 3′ direction (see the entries "Nucleotides" and "3′-end and 5′-end").

DNA sequencing This is the process of determining (or "reading") the order of nucleotide bases in a DNA molecule of interest. It can be done by a variety of techniques, and the recent advent of rapid DNA sequencing methods has greatly accelerated biological and medical research and discovery.

FRET (Förster resonance energy transfer or fluorescence resonance energy transfer) It is a mechanism describing nonradiative energy transfer between two light-sensitive molecules (chromophores or fluorophores). In certain cases, one of these molecules acts as a quencher, which greatly reduces fluorescence of another molecule when they are in close proximity to each other.

Gel electrophoresis This is a laboratory method used to separate mixtures of DNA fragments according to their size by driving them with an electric field through a gel that contains small pores.

Hybridization or DNA hybridization This is the process of non-covalently joining two complementary single strands of DNA into a DNA duplex via Watson-Crick pairing between A, G, C, and T bases in one strand to, respectively, T, C, G, and A bases in another strand. The hybridization duplexes may be dissociated by thermal denaturation, also referred to as melting. In the absence of external negative factors, the processes of hybridization and melting may be repeated in succession indefinitely, which lays the ground for polymerase chain reaction (PCR) and DNA-based molecular machines fueled by DNA.

Hydrogen bond This is a weak non-covalent bond between two molecules resulting from an electrostatic attraction between a proton in one molecule and an electronegative atom in the other.

Intercalation It is the insertion of small molecules, called intercalators, between the planar bases of DNA. Most of the DNA intercalators are polycyclic, aromatic, and planar molecules.

Kb (kilobase or kilobase pairs)   A unit of measurement in molecular biology equal to 1,000 base pairs of duplex DNA or RNA.

Messenger RNAs (mRNAs)   These are a group of specific nucleic acid molecules, which convey genetic information from DNA to proteins. As in DNA, mRNA genetic information is in the sequence of nucleotides, which are arranged into codons consisting of three base pairs each. Each codon encodes for a specific amino acid, except the stop codons, which terminate protein synthesis.

Micrometer (aka micron)   It is one-millionth of a meter and can also be expressed as 10^{-6} m.

Nanometer (nm)   A billionth part of a meter; 1 nm = 10^{-9} m.

Non-covalent bonds   These differ from a covalent bond in that they do not involve the sharing of electrons, but are based instead on less strong electrostatic and hydrophobic interactions.

Nucleotides   These are the individual structural units of the polynucleic acids, including DNA. A nucleotide is composed of a nucleobase (nitrogenous base of four different kinds abbreviated as A, T, G, or C), a five-carbon sugar ($2'$-deoxyribose in DNA), and a phosphate group. Without the phosphate group, the nucleobase and sugar compose a nucleoside. A nucleotide can thus also be called a nucleoside phosphate.

Oligonucleotide   It is a short-length DNA single strand, typically with hundred or less nucleotide units. Commonly, they are chemically assembled in a sequence-specific manner from the four modified monomeric units corresponding to A, T, G, and C nucleotides. Oligonucleotides are able to readily bind to their respective complementary DNA sequences—the property that is employed in their numerous uses, such as PCR primers and ligation linkers.

One-pot reaction (aka one-pot synthesis)   It is the production of a new chemical structure involving multiple steps and several reaction sequences designed in such a way that it could proceed within one single container.

Overhangs and sticky ends   See Fig. 3.2II in this book.

Parallel strands   DNA strands are parallel if they run alongside each other in the same direction (see the entry "$3'$-end and $5'$-end" above).

PCR or polymerase chain reaction   It is a powerful technique to amplify a few copies of a piece of DNA by million to billion times in just a few hours, thus generating an enormous number of replicas of a particular DNA sequence. The method relies on thermal cycling, consisting of cycles of repeated heating and cooling of the reaction mix for separating (melting) the complementary DNA single strands from DNA duplex and their enzymatic replication. As PCR progresses, the newly generated DNA strands are themselves used as additional templates for replication, therefore establishing a chain reaction in which the DNA template is exponentially amplified (also see "DNA polymerase" and "Primer").

Peptide nucleic acid (PNA)   It is an artificially synthesized oligomer, which consists of non-charged protein-like backbone with attached nucleobases. PNA oligomers are the DNA mimics in a sense that they are of different chemical structure, but could bind complementary DNA oligomers.

Primer (or priming oligonucleotide)   This is a short single strand of DNA (aka oligonucleotide) that hybridizes to a longer template DNA strand to serve as a starting point for the synthesis of complementary DNA strand. They are required for DNA replication because DNA polymerases can only add new nucleotides to an existing strand of DNA. The polymerase starts replication at the 3′-end of the primer and copies the opposite DNA strand by extending primer in a primer extension reaction, aka primer elongation reaction (also see the entry "3′-end and 5′-end").

Purines   These are a distinct class of the two-ringed chemical compounds that includes DNA nucleobases adenine and guanine.

Pyrimidines   These are a distinct class of the single-ring chemical compounds that includes DNA nucleobases cytosine and thymine.

Recombinant DNA   It is an artificial DNA molecule created by so-called genetic engineering techniques from segments of different natural DNAs to produce recombinant genes and hence recombinant proteins.

Replication   Copying or reproducing genetic material or artificial DNA and any nano-construct made of it; each act of replication results in doubling of replicated material.

Restriction nucleases (restriction enzymes)   These are the enzymes that cut DNA strands at a specific sequence of nucleobases. For example, EcoRI restriction nuclease will cut DNA strand only after the G and only in the sequence GAATTC, whereas HindIII nuclease will cut DNA strand only after the A and only in the sequence AAGCTT. Both EcoRI and HindIII digestions produce sticky ends.

RNA   Acronym for ribonucleic acid, which is a biopolymer akin DNA in its chemical structure, but it carries a different sugar in a backbone—a ribose, instead of deoxyribose in DNA. Like DNA, RNA can also form duplex structures (involving either RNA or DNA counterparts); however in nature most RNA molecules exist as single strands.

Secondary structure of DNA   It is the local 3D structure of helices, loops, and junctions adopted by a polynucleotide chain due to interactions between neighboring residues. These include DNA hairpins (stem-loops) and DNA cruciforms.

Self-assembly   This is the process by which certain organic and inorganic molecules adopt a defined arrangement without guidance or management from an outside source; there are two types of self-assembly: intramolecular self-assembly and intermolecular self-assembly.

Semiconductor   This is a solid substance that has an electrical conductivity between that of a metal, like copper and gold, and an insulator, such as glass. Its conductivity increases with rise of the temperature, which is opposite to that of a metal. Semiconducting properties of a material can be altered in useful ways by the deliberate, controlled introduction of specific impurities ("doping") into the crystal structure. Devices made of semiconductors, notably silicon, are essential components of most electronic circuits.

Sticky end   This is an end of a DNA double helix at which a few unpaired nucleotides of one strand extend beyond the other. This name emphasizes the fact that a complementary pair of such dangling nucleotide strands will stick together, as it is shown in Fig. 3.2 of this book, for example.

Strand displacement (aka strand exchange)   It is a process when one strand in a double-stranded DNA complex is gradually rejected from DNA duplex and replaced by a third DNA strand, usually referred to as the invading strand, which is complementary to the remaining DNA strand and therefore forms another DNA alternative duplex.

Streptavidin   It is a relatively small tetrameric protein with four identical subunits purified from the bacterium *Streptomyces avidinii* (here is the name). It has an extraordinarily high affinity for biotin, which is one of the strongest non-covalent interactions known in nature. Streptavidin is used extensively in molecular biology and bio-nanotechnology due to the extreme resistance of biotin-streptavidin complex to various harsh conditions, including high temperatures and extreme pH values.

Transistor   A semiconductor device with three connections commonly used to amplify or switch electronic signals.

Z-DNA   A left-handed double-helical form of DNA in which the double helix winds to the left in a zigzag pattern instead of winding to the right and in a smooth manner, as in the more common B-DNA form. B-DNA can be converted into Z-DNA by self-rewinding the double helix from right to left if DNA molecule has a Z-DNA-favoring sequence and/or is in the Z-DNA-favorable environment, such as high-salt conditions and special cations.

Index

Printed in the United States
By Bookmasters